# Innovative Materials and Techniques for Osteochondral Repair

Horea Rareș Ciprian BENEA, MD, PhD

Published by **Materials Research Forum LLC**
Millersville, PA 17551, USA

Published as part of the book series
**Materials Research Foundations**
Volume 62 (2019)
ISSN 2471-8890 (Print)
ISSN 2471-8904 (Online)

Print ISBN 978-1-64490-052-9
ePDF ISBN 978-1-64490-053-6

Distributed worldwide by

**Materials Research Forum LLC**
105 Springdale Lane
Millersville, PA 17551
USA
http://www.mrforum.com

Printed in the United States of America
10 9 8 7 6 5 4 3 2 1

# Table of Contents

# Preface

This study will bring real scientific benefits, from the promise of these stem cell therapies, which is a hope for the future in the treatment of focal joint cartilage defects and not only, examples that can lead to the treatment of arthrosis, delays in consolidation/non-unions, osteoarticular infections, etc.

The study will help deepen the understanding of mesenchymal stem cells role in the treatment of articular cartilage lesions and will enable the improvement of interventional treatment options by developing methods with minimal cost and morbidity and maximum efficacy. It will also serve as the basis for algorithms for the diagnosis and treatment of cartilage lesions in current practice. The pilot study on testing the feasibility of using lipoaspirate fluid (LAF) cell is a precursor to a broader study regarding the testing of the use of these cells in the treatment of arthrosis, and the findings of which should be more relevant.

If other results demonstrate it, the fluid fraction of lipoaspirate may be further used as a source of mesenchymal stem cells with applications in multiple fields: orthopedics, general surgery, plastic and repair or vascular surgery, having the advantage of relative simplicity and rapidity of the separation method of lipoaspirate fluid from the lipoaspirate.

The innovative elements of the research are represented by:

- LAF separation process, consisting of stem cells and stimulating nutrients;
- using LAF cells for chondral repair;
- repair technique using stem cells seeded on the collagen I/III membrane (capable of being used and developed in a "one step procedure");
- the use for the first time of demineralized MatriBone Ortho collagen I/III membrane;
- direct comparison of the two types of stem cells in terms of the results obtained;
- creation of the virtual mathematical pattern based on the CT imaging of the sheep knee, which can be used for finite element analysis;
- MRI with the 7 Tesla machine revealing structural details of microscopic clarity, demonstrating its potential for use to explore the structure of articular cartilage and subchondral bone;

- introducing cartilage tissue regeneration therapies in the current clinical practice using collagen scaffolds and multipotent mesenchymal cells, which offers the possibility of making real "biological arthroplasties";
- obtaining encouraging results on the regeneration of cartilage tissue by injecting stem cell concentrates, this gives hope to patients suffering from cartilage degradation.

*Horea Rareş Ciprian BENEA, MD, PhD*
*Lecturer at "Iuliu Hatieganu" University of Medicine and Pharmacy,*
*Cluj-Napoca, Romania.*

# List of Abbreviations

| | |
|---|---|
| **2-ΔΔCt** | Analysis of Relative Gene Expression Method 2-ΔΔCt |
| **ACAN** | Aggrecan |
| **ACI** | Autologous Chondrocyte Implantation |
| **ACL** | Anterior cruciate ligament |
| **ADAMTS-4** | Disintegrin and metalloproteinase with thrombospondin motifs 4 |
| **ADAMTS-5** | Disintegrin and metalloproteinase with thrombospondin motifs 5 |
| **AFM** | Atomic Force Microscopy |
| **AMIC** | Autologous Matrix-Induced Chondrogenesis |
| **AOFAS** | American Orthopaedic Foot and Ankle Society |
| **ASC** | Adipose-derived Stem Cells |
| **BMC** | Bone Marrow Concentrate |
| **BMP** | Bone morphogenetic protein |
| **BMS** | Bone Marrow Stimulation |
| **CAD** | Computer-aided design |
| **CD105+** | Cluster of Differentiation 105+ |
| **CD44** | Cluster of Differentiation 44 |
| **CFU-F** | Colony-forming unit fibroblasts |
| **COL2A1** | Collagen type II alpha 1 chain |
| **COMP** | Cartilage oligomeric matrix protein |
| **CORS** | Consistent Osteochondral Repair System |
| **CS846** | Chondroitin Sulfate epitope 846 |
| **CT** | Computed tomography |
| **CTX-II** | C-terminal telopeptides of type II collagen |
| **DAPI** | 4',6-Diamidino-2-Phenylindole |
| **dGEMRIC** | Delayed gadolinium-enhanced MRI of cartilage |
| **DMEM** | Dulbecco's Modified Eagle's Medium |
| **DMEM/F12** | Dulbecco's Modified Eagle Medium/Nutrient Mixture F-12 |
| **DS** | Standard deviation |
| **ECM** | Extracellular matrix (of cartilage) |
| **EDTA** | Ethylenediaminetetraacetic acid |
| **FBS** | Fetal bovine serum |
| **FEA** | Finite element analysis |
| **FGF** | Fibroblast growth factor |
| **FLASH** | Fast Low Angle Shot |
| **FSE** | Fast Spin Echo |
| **GAG** | Glycosaminoglycans |

| | |
|---|---|
| **GAPDH** | Glyceraldehide 3-phosphate dehydrogenase |
| **HAP** | Hydroxyapatite |
| **ICRS** | International Cartilage Repair Society |
| **IGF-1** | Insulin-like growth factor 1 |
| **IGF-1 BPs** | Insulin-like growth factor 1 binding protein |
| **IKDC** | International Knee Documentation Committee |
| **IL-1β** | Interleukin 1 beta |
| **IL-6** | Interleukin 6 |
| **IL-8** | Interleukin 8 |
| **ILGF** | Insulin-like growth factor |
| **ILK-1** | Integrin linked kinase 1 |
| **LAF** | Lipoaspirate fluid |
| **LTB4** | Leukotriene B4 |
| **MACI** | Matrix-induced Autologus Chondrocyte Implantation |
| **MMPs** | Matrix metalloproteinases |
| **MPFL** | Medial patello-femoral ligament |
| **MRI** | Magnetic resonance imaging |
| **MSC** | Mesenchymal stem cell |
| **NEA** | Non-essential amino acids |
| **OATS** | Osteochondral Autograft Transfer System |
| **OCD** | Osteochondritis dissecans |
| **PBS** | Phosphate-buffered saline |
| **PCL-g-COL-g-CS** | Poly($\varepsilon$-caprolactone)-graft-type II collagen-graft-chondroitin sulfate |
| **PDGF** | Platelet-derived growth factor |
| **PG** | Proteoglycans |
| **PGE2** | Prostaglandin E2 |
| **PLA** | Processed lipoaspirate |
| **PLGA** | Porous poly(DL-lactic-co-glycolic acid) |
| **PRP** | Platelet-rich plasma |
| **qPCR** | Quantitative polymerase chain reaction |
| **RARE** | Rapid Acquisition with Relaxation Enhancement |
| **ROS** | Reactive oxygen species |
| **SE** | Standard error |
| **SEM** | Scanning electron microscope |
| **SIFK** | Subchondral Insufficiency Fracture of the Knee |
| **SLRP** | Small leucine-rich proteoglycan |
| **sMSCs** | Sheep Mesenchymal Stem Cells |
| **SONK** | Spontaneous Osteonecrosis of the Knee |
| **SOX-9** | Sex determining region Y-box 9 |
| **SVF** | Stromal Vascular Fraction |
| **T1 rho** | T1 relaxation time in the rotating frame |

| | |
|---|---|
| **TCP** | Tricalcium phosphate |
| **TGF-β** | Transforming growth factor beta |
| **TNF-α** | Tumor necrosis factor alpha |
| **TRITC** | Tetramethylrhodamine isothiocyanate |
| **uPA** | Urokinase-type plasminogen activator |
| **WOMAC** | The Western Ontario and McMaster Universities Osteoarthritis Index |
| **β-TCP** | Beta-tricalcium phosphate |

# Introduction

At present, the pathology and treatment of cartilage lesions represents a field of intensive research, both at the laboratory and clinical level. Restoring articular cartilage through hyaline tissue formation is still a challenge both for surgeons and researchers. There are no broad-batch clinical or lab studies to establish objective conclusions about the best therapeutic methods.

The morphological and metabolic characters of cartilaginous tissue have prompted a lively interest among both clinicians and researchers. The failure of the classical methods of treatment has led to the search for biological solutions in the attempt to heal injuries. There is an increasing focus on the necessity of restoring the cellular capital and not only the fundamental substance. The potential of stem cells development towards the cartilaginous line is recognized, but also towards the recovery of the subchondral bone, the essential element of support and nutrition for the cartilage. Direct use of cartilage cells is also attempted to proliferate locally at the defect level under the influence of different biological stimuli without the need for laboratory cultivation.

Moreover, the development of implants with biological role in the repair of osteochondral tissue, support scaffolds capable of receiving mesenchymal stem cell elements (multipotent somatic precursor cells located in the perivascular areas of the connective stroma of adult tissues) to be directed under the influence of biological factors, cytokines, specific growth factors, towards the recovery of cartilaginous and bone cells is more and more tested.

These therapies bring us already in the field of regenerative medicine, which is an area of intensive development that makes an important contribution to the medical sciences. Here, the tissue engineering of bio composite materials has found a breeding place for an extensive development. Along with biomaterials, stem cells are key points of regenerative medicine, and their ability to differentiate and renew makes them essential for tissue repair and organ regeneration in the living organism.

The results are encouraging, showing the enormous possibilities of research this field offers, but in regard with the therapeutic methods, there is still necessary the development of broad experimental studies on animal models and their clinical applicability evaluation. The goal is to develop simple, reproducible methods, with minimal morbidity and risks, but cheaper, with better results and less complications. All the researchers focused on this domain should agree in order to homogenize the procedures and results.

Most procedures based on the use of mesenchymal stem cells are time-consuming, technically difficult and require multiple interventions and *ex vivo* handling, thus involving high costs, risk of contamination and local infections. The ideal procedure should be a single-step with minimal tissue manipulation. In this regard, cells derived from bone marrow (BMCs) and adipose tissue (ASCs) are a kept promise, as we will further on see.

I have decided to develop the research in a field of topical science, namely the stem cell autologous biological therapies with potential to be used to restore articular cartilage injuries. These therapies are already used in current practice and encouraging results have been reported.

I planned to develop a therapeutic method, as simple as possible from a technical point of view, as fast and less aggressive, and as effective as possible in the treatment of articular osteochondral lesions.

Achieving good results through an animal model study would create the premise of validating this "one-step surgery". This would represent a true novelty in the field of articular cartilage surgery and would have every chance of successful implementation in current practice, being practically the goal of various research that is currently taking place worldwide. Furthermore, the population impact would be huge, given that joint pain affects millions of people.

In this respect, I have designed five studies, the objectives of which being as follows:

- Assessing the possibilities of applying stem cell based treatments, both in animals and in clinic
- Assessing the behavior of stem cells on certain materials that could be used as scaffolds
- Testing the feasibility and applicability of stem cell chondral repair procedures
- Comparative evaluation of the results of these therapies by methods of histological, molecular, medical imaging investigation
- Testing new materials and procedures for cellular sampling
- Developing virtual mathematical patterns for finite element analysis
- Assessing the incidence and outcome of the chondral pathology in patients treated by arthroscopic methods
- Translating therapeutic procedures from the laboratory, on animals and then to the human medicine clinic.

## 1.    Normal articular cartilage

The articular cartilage represents a thin layer of specialized conjunctive tissue with specific mechanical and biological properties[1]. Its main function is to ensure a smooth lubricated surface that reduces articular friction and facilitates transmission of forces to the subchondral bone. The cartilage is unique by its capacity to resist to cyclic loading, without suffering major degenerative changes or damage[2].

### 1.1    Anatomy and histology

### 1.1.1    Generalities

The articular cartilage is a specialized conjunctive tissue situated at the level of articular surfaces. The main characteristics of it are the lack of innervation, blood and lymphatic supply and the presence of a single type of cells, the chondrocytes[3]. The role of these cells is the synthesis and remodeling of the extracellular matrix, rich in collagen fibers and glycosaminoglycans (GAG) that takes part in forming the proteoglycans (PG)[3].

The nutrition of articular cartilage is realized exclusively by diffusion from synovial liquid and subchondral bone, facilitating in the same time the elimination of metabolic wastes. The diffusion of nutrients depends on the size (maximal molecular weight of 70 kDa), shape and electrical charge of molecules, and also on the proteoglycans concentration in the cartilage[4].

Due to the composition and distribution of collagen fibers, the articular cartilage possesses special biomechanical properties, by the capacity of amortization and repartition of mechanical forces that act on it, leading to the reduction of forces applied on the joints during their activity. The articular cartilage is a stable form of cartilage, resisting to vascular invasion, mineralization and replacement with bony tissue, thus differentiating of growth plates cartilage, which is transitory and ossifies by the end of puberty[5].

### 1.1.2    Types of cartilage

   a) The hyaline cartilage is white, translucent and vitreous, made of collagen fibers, most of them type II, and proteoglycans that confers a high refractory index. This type of cartilage is mainly localized at the level of the articular surfaces, allowing the bone parts to glide, the absorption and distribution of compression and friction forces that appear during joint motion. The hyaline cartilage can also be identified at the level of the nasal septum, chondro-sternal joint, larynx, tracheas and bronchi. Also during fetal stage, the skeleton is made of hyaline cartilage that will permit the development and growth of long bones by enchondral ossification processes[6].

b) The elastic cartilage differentiates from the hyaline by its abundance of elastin fibers that offers elastic properties and a high degree of flexibility. This type is localized at the level of the external ear, nasal wings and epiglote[7].

c) The fibrous cartilage, apart from the hyaline, is only made of type I collagen fibers that offers increased resistance to compression and tension forces. It is localized at the level of insertion of ligaments and tendons, meniscus, pubic symphysis and in the fibrous ring of the intervertebral discus[8].

### 1.1.3 Articular cartilage composition

### 1.1.3.1 Chondrocytes

The chondrocyte cells represent approximately 10% of total cartilaginous tissue volume (Figure 1.1). These are highly specialized adult cells, metabolic active, playing a unique role in the development, maintenance and regeneration of extracellular matrix.

The chondrocytes have embryonic origin from the mesodermal layer for limbs elements and from the ectodermal layer (neural ridges) for the facial skeleton[9]. They present a round or polygonal morphology, but they can also be found as discs plates, function of their localization in the articular cartilage layers. In their cytoplasm numerous lysosomes are identified, as well as lipid vacuoles and glycogen.

Taking into account the fact that chondrocytes nourish by diffusion, their metabolism is anaerobe, thus assuring the synthesis and regeneration of extracellular matrix. The chondrocytes produce macromolecules that are present in the extracellular matrix (procollagen, proteoglycans and hyaluronic acid) and also the enzymes responsible for their degradation, like cathepsins (type B and L), metalloproteinases (collagenases) and agreccanases[10].

Each chondrocyte produces to itself a specialized microenvironment and it is responsible for surrounding extracellular matrix turnover. This microenvironment traps chondrocytes in their own matrix, and thus prevents their migration in the surrounding cartilage areas. Seldom, the chondrocytes are forming intercellular connections for direct transduction of the signal and intercellular communication. In exchange they respond to a variety of stimuli like growth-factors, mechanical loading, piezoelectric forces and hydrostatic pressure. Their role is to synthetize glycosaminoglycans and type II collagen. The synthesis of sulphate glycosaminoglycans is hormonally mediated: the growth hormone, thyroxine and testosterone increase the synthesis, while cortisol is decreasing it[11]. The chondrocytes have a limited replication potential, fact that limits the intrinsic healing capacity of the cartilage in response to a lesion. The chondrocyte life-span is variable, depending on the existence of an optimal environment in regard to biochemical and mechanical properties.

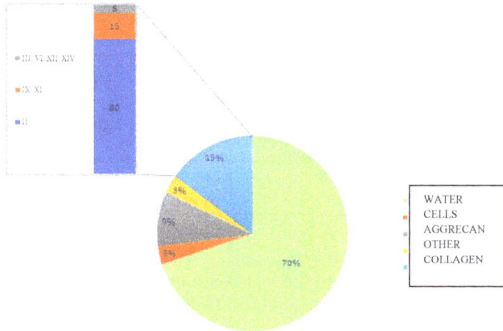

***Fig. 1.1.*** *Molecular composition of articular cartilage.*

### 1.1.3.2 The extracellular matrix of the articular cartilage

The extracellular matrix of the articular cartilage (ECM) is made of a collagen fibers network that constitute fibrillary scaffolds, disposed in a fundamental substance composed mainly of proteoglycans and water. The collagen fibers located at the ECM are mainly of type II, IX and XI, but other non-collagen proteins are encountered, like COMP (cartilage oligomeric matrix protein) that facilitates the firm and favorable attachment of the lubricin, with role in friction forces reduction[12], sometimes small PG, rich in leucine (SLRP), surface proteoglycans, hyaluronic acid and chondroitin sulphate. The extracellular matrix plays an essential role in the regulation of metabolism and chondrocyte function, by organizing the cytoskeleton[13].

*The collagen from articular cartilage*

At the level of articular cartilage one can identify twelve types of collagens: types I, II, III, V, VI, IX, X, XI, XII, XIII, XIV and XVI.

The collagen fibers represent about 60% of total solid cartilage components weight. Type II collagen fibers are predominant (80-85% of all collagen fibers from ECM) (Table 1.1.). The types I, IV, V, VI, IX and XI of collagen are present in a small proportion and they play the role in formation and stabilization of the type II collagen network[14]. Type II collagen fibers are differently oriented: superficially they are parallel with cartilage surface, in the middle part they are oriented obliquely and in profundity they are perpendicular to the bony surface. The proteoglycans aggregates are bonded to the collagen fibers, forming a network in which there are no water nor electrolytes[15].

***Table 1.1.*** *The main types of collagen at the level of articular cartilage, and their function*[16]

| Collagen type | Morphologic-localization | Function |
|---|---|---|
| II | Principal macro-fibrillar component | Resistance against traction forces |
| VI | Peri-cellular matrix | Facilitate chondrocyte adherence to the matrix |
| IX | Cross-link bonding to the surface of macrofibrils | Inter-fibrillary connections |
| X | Close connection to the hypertrophic chondrocytes from calcified layer | Structural support Helps mineralizing the cartilage |
| XI | Associated or no to macrofibrils | Formation of the nucleate fibrils |

There are 15 different types of collagen, made of at least 29 polypeptide chains. All the collagen family members contain a region made of 3 polypeptide chains arranged in a triple helix. The greatest proportion of amino acids in polypeptide chains composition is first represented by glycine and proline. Hydroxyproline assure the stability through hydrogen bonds disposed along the molecule. The polypeptide chains disposed as triple helix offer the articular cartilage important properties against shearing and traction forces, contributing to matrix stabilization[17].

Type II collagen is the main structural constituent of the articular cartilage, representing about 25% of dried weight of the extracellular matrix (ECM) and 80% of total collagen quantity in articular cartilage. This type of collagen is also found in other tissues, like intervertebral discs, retina, cornea and vitreous body. Type II collagen is codified by COL2A1 gene and the product is a homotrimer made of 3 $\alpha 1$[18] chains. The mutations at the COL2A1 gene are responsible of affections like achondrogenesis, hypochondrogenesis, dysplasia or some forms of familial osteoarthritis[19].

*The proteoglycans in articular cartilage*

Proteoglycans are strongly glycosylated monomers of proteins. In articular cartilage, they represent the second largest group of macromolecules from extracellular matrix and they weigh about 10-15% of wet weight. Proteoglycans are formed of a protein core and more linear chains of glycosaminoglycans attached by covalent binding. The chains are made of more than 100 monosaccharides and they extend outside the protein core, remaining separated one from the other due to charge repulsion force[20].

The articular cartilage contains a great variety of proteoglycans that are essential for its normal function, including aggrecan, decorin, biglycan and fibro-modulin. Aggrecan is the biggest and heaviest proteoglycan, containing more than 100 chondroitin sulphate chains and 50 of keratin sulphate[21]. Aggrecan is characterized by its capacity to interact with hyaluronic acid and to form great proteoglycan aggregates with the help of binding proteins. This proteoglycan occupies the interfibrillar space of the ECM and offers the cartilage essential osmotic properties for its capacity to resist compression forces[22].

Proteoglycans that are not aggregated have the capacity to interact with collagen. Although decorin, biglycan and fibro-modulin are much smaller than the aggrecan, they can be found in similar molar quantities. The structure of these molecules is strongly tied to that of proteins, so they differ by the aminoglycans components and by the function deserved. Decorin and biglycan contain 1 and respectively 2 chains of dermatan sulphate, while fibro-modulin contains more chains of keratansulphate[23]. Decorin and fibro-modulin interact with type II collagen fibers and they play a role in fibrils genesis and interfibrillar interactions[24]. Biglycan is mainly found next to chondrocytes, where it could interact with type IV collagen fibers[25].

Although the non-collagenous proteins and glycoproteins are found at the level of articular cartilage, their function was not completely characterized. Some of these molecules (like fibronectin and CII-anchorin from the surface of chondrocytes) play an important role in organizing and maintenance of extracellular matrix structure[26].

Other constituents of the articular cartilage are: integrin, fibronectin, thrombospondin, lubricin or annexin V.

*Water*

The most impotent element from the articular cartilage composition is water, represented in a proportion of about 70-80%. Approximatively 30% of it is found in the intrafibrillar space, a small percent in intracellular and the rest in the pores of the matrix. Inorganic ions, sodium, calcium, chloride, potassium are dissolved in the tisular water[27]. The relative water concentration decreases from about 80% in the superficial zone to approximately 65% in profundity. The water flow in the cartilage and at the articular surface helps for transportation and distribution of nutrients at the level of chondrocytes and also lubricates the articular surfaces[28].

Most of the interfibrillar water is found in gelatinous state and it can be mobilized through the ECM if a pressure gradient is applied by compressing the tissue and subsequently the solid matrix. The frictional resistance against this flow is very high and thus the tisular permeability is very low[29]. The property of cartilage to resist significant loading is due to a combination between the friction resistance of the water flow and water pressure inside the matrix.

### 1.1.4 Histology of articular cartilage

The ultrastructure of collagen fibers, ECM and chondrocytes contribute to the division of articular cartilage in 4 distinct histologic zones or layers: superficial, middle, profound and calcified (Fig. 1.2). At their turn, these 3 zones are subdivided in 3 regions: pericellular, territorial and inter-territorial.

1. The superficial or tangential zone represents about 10-20% of cartilage thickness and protects the profound layers from the stress caused by shear forces. It is situated in direct contact with the synovial liquid. It is responsible of cartilage resistance to traction and compression developed in the joints. Collagen fibers (mainly type II and IX) are tightly packed and disposed parallel to the articular surface. This layer contains a great number of flattened chondrocytes[29]. Its integrity is important for protection and maintenance of profound layers.

2. The middle or transition layer is situated immediately under the superficial zone and represents a region of anatomic and functional transition from the superficial to the profound zone. It represent 40-60% of total cartilage volume. It contains proteoglycans and thick collagen fibers disposed obliquely. The chondrocytes are relatively low-numbered and they have spherical shape. From the functional point of view, this layer is the first line of resistance against the compression forces[30].

3. The profound zone contains collagen fibers disposed perpendicularly to the cartilage surface, offering a great resistance to compression forces. This zone is made of collagen fibers with the greatest diameter, disposed radially, with the highest content of proteoglycans and the lowest water concentration. The chondrocytes are disposed in columns, parallel to the collagen fibers and perpendicular to the articular line. The profound zone represents about 33% of cartilage volume, having the function to offer resistance against compression forces, due to the increased content in proteoglycans and the radial disposition of collagen fibers[31].

4. The calcified layer plays an essential role in cartilage bonding to the bone, anchoring the profound collagen fibers to the subchondral bone. In this zone, the cell population is reduced and the chondrocytes are hypertrophic[32].

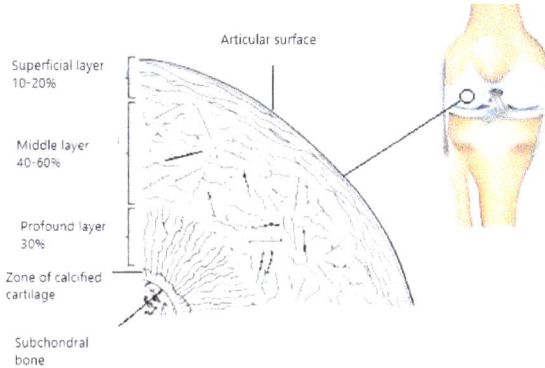

**Fig. 1.2.** *The orientation of collagen fibers in articular cartilage layers.*

The matrix is formed of several regions differentiated by the distance between chondrocytes, composition, collagen fibers diameter and organization. Thus, the ECM can be divided into three histologic regions:

a) The peri-cellular region is represented by a thin layer adjacent to the cellular membranes that completely surrounds the chondrocytes. It contains mainly proteoglycans, glycoproteins and other non-collagenous proteins. This region could play a functional role in the initiation of transduction signal for cartilage synthesis in the portent regions[33].

b) The territorial region surrounds the peri-cellular matrix and is composed of thin collagen fibers that form a network around cells. This region is thicker than the peri-cellular matrix and it is supposed to play a role in cartilage cells protection from the mechanical stress and to contribute to cartilage elasticity and capacity to resist substantial loading[34].

c) The inter-territorial region is the best represented. It contributes the most to biomechanical properties of articular cartilage. It is characterized by random orientation of thick collagen fascicules, disposed parallel to the articular surface in the superficial zone, oblique in the middle zone and perpendicular in the profound zone. This region contains an increased quantity of proteoglycans[25].

## 1.2     Physiology of cartilaginous tissue

### 1.2.1   Chondrogenesis

Starting from embryonic period, the cartilage is the first tissue that ensures the skeletal support of the body, being generated by chondrogenesis process. At the level of growth plates it plays an important role in the enchondral ossification process, being progressively replaced by bone tissue, meanwhile at the level of articular cartilage it is permanent and does not calcify. Chondrogenesis is initiated by the differentiation of cells from ectodermal neural crests (for the cranial-facial skeleton) and cells from mesodermal somatopleura from the lateral plate (peripheral skeleton)[35].

### 1.2.2   Metabolism

In adults, the articular cartilage matrix is separated from subchondral vascular space by the subchondral plate[36]. Articular cartilage nutrition is done by diffusion from the synovial liquid. The cartilaginous matrix limits the passage of materials in function of dimension, electrical loading and molecular configuration.

The mean dimension of pores in the ECM is about 6.0 nm. Without a direct source of nutrients from blood of lymphatic vessels, chondrocytes depend on anaerobic metabolism, first of all. Chondrocytes are responsible for the development, maintenance and reparation of ECM through a group of degradation enzymes. Chondrocytes synthesize components of the ECM like proteins and lateral catenae of glycosaminoglycans. The metabolic activity of chondrocytes can be altered by a great variety of chemical and mechanical factors, like pro-inflammatory cytokines (ILK-1 and TNF-α)[37], with catabolic and anabolic effects, that play an important role in degradation and synthesis of ECM molecules. Proteoglycans are synthesized, maintained and released in the ECM by chondrocytes.

A series of growth factors and regulating peptides are involved in proteoglycan metabolism regulation, among which are ILGF, TGF- β, ILK-1 and TNF−α[38]. Very little is known about the molecular mechanism through which these growth factors and peptides exercise their effects on proteoglycan metabolism. Chondrocytes are protected from potential harming biomechanical forces by the surrounding ECM. The homeostasis of ECM metabolism is maintained by a balance between macromolecules degradation and their replacement with the new synthesized ones[39]. Proteoglycans turn-over can last up to 25 years[40], while collagen half-life time can vary between decades up to 400 years.

The most important proteins implied in cartilage turn-over are metalloproteinases (like collagenase, gelatinase and stromelysin) and cathepsins (B and D cathepsins)[38]. The collagenases degrade denatured type II and IV collagen fibers and they also act on fibronectin, elastin and type V, VII, X and XI collagen. The role of stromelysin is to degrade the core of aggrecans. All metalloproteinases are secreted as inactive proenzymes that need extracellular environment for activation[41].

Load-bearing and mobility of a joint are important for normal maintenance of articular cartilage structure and function, repeated movement and dynamic loading being essential for local metabolism maintenance. It was shown that inactivity of a joint leads to cartilage degradation, and the development of affections like osteoarthritis is associated with dramatic changes in articular cartilage metabolism. This can happen when there is an unbalance between degradation and synthesis[34].

### 1.2.3   Mechanical properties of articular cartilage

The biomechanical behavior of cartilage is better understood when this tissue is regarded as a biphasic medium, with a fluid and a solid phase. Water is the principal component of liquid phase, representing up to 80% of tisular wet weight. Inorganic ions like sodium, calcium, chloride and potassium are also identified in the fluid phase. The solid phase is represented by the ECM, which is porous and permeable.

The relationship between proteoglycans aggregates and interstitial fluid ensures the compression resistance of the cartilage by negative electrostatic repulsion forces. The rapid initial application of contact forces during articular loading determines an immediate increase of interstitial fluid pressure. This local pressure elevation determines the fluid to exit the ECM, generating an increased friction force at this level. When the compression force is removed, interstitial fluid comes back in the tissue. The low permeability of articular cartilage stands against the sudden exit of fluid from the matrix[42].

The articular cartilage is viscoelastic, existing two types of mechanisms that are responsible for this unique characteristic: a flux-dependent mechanism and another one flux-independent. The dependent one is influenced by interstitial fluid and friction forces associated with this, while the flux-independent mechanism is due to the specific macromolecules movements[43]. As consequence, the fluid pressure ensures a significant component responsible for total load support, and thus reducing the stress acting on the solid matrix[41].

The articular cartilage presents fluctuant deformation and stress and relaxation response. When a constant stress is applied on the tissue, its deformation increases in time and it will modify until an equilibrium value is reached. Similarly, when cartilage is deformed and kept under a constant tension, the stress will reach a maximum, followed by a slow process of stress-relaxation, until the equilibrium point is reached[2].

The complex composition and organization of the cartilage in the middle layers contributes significantly to its resistance to shearing forces. The elongation of randomly disposed collagen fibers determines a stress response to shearing forces. The property of resistance to traction derives from the arrangement of molecules in the collagen fibers. The fibers stability and resistance to tearing by traction results from the intra and intermolecular crossed links.

### 1.2.4   Extracellular matrix remodeling

The cartilaginous extracellular matrix remodeling and replacement is ensured by chondrocytes that synthesize the enzymatic components with role in macromolecules degradation (collagen, proteoglycans), their activity being controlled by physical, chemical and mechanical stimuli. Growth-factors like PDGF, FGF, TGF-β and BMP[44], altogether some specific hormones (growth hormone, calcitonin, steroids, etc.)[45] stimulate chondrocyte proliferation and cartilaginous matrix components synthesis.

## 2.    Physiopathology of cartilage lesions

Damage of the articular cartilage can have multiple causes, being the result of acute traumatic lesions, early posttraumatic degenerative changes, repetitive micro traumatic events, articular development issues, as well as gained metabolic factors[46-49].

Articular cartilage response to lesions and its evolution is yet another controversy. According to the data described by Widuchowski et al., a singular lesion is not incriminated to statistically significant increase the further development of arthritic changes[50]. On the other hand, according to imagistic examinations, it was pointed out that articular cartilage lesions lead to an accelerated chondral deterioration[51-54], the coexistence of these lesions with anterior cruciate ligament lesions leading to significantly lower functional scores compared to the patients who only present a ligament lesion[55].

### 2.1    Response of chondral tissue to lesions – spontaneous repair

Articular cartilage is a metabolically active structure, but present a reduced healing capacity, due to the relative absence of young non-differentiated cellular populations and lack of vascular supply[56]. Mechanical resistance depends on the integrity of the extracellular matrix composed mainly of collagen type II and proteoglycans[57]. Because of these specifications, the response of the articular cartilage to deep and superficial structures is different[58].

For example, during the natural evolution of deep osteochondral lesions, the differentiation of mesenchymal cells within the bone marrow leads to the development of hyaline cartilage in the first 4 to 8 weeks after lesion[58]. The new cartilage does not present the structure and composition of the native cartilage, as early signs of chondral deterioration can be seen within 12-24 weeks[58]. The new tissue is an intermediate between hyaline tissue and fibrocartilaginous tissue and does not produce a complete integration, as the collagen fibers do not interconnect with the adjacent cartilage, the latter having necrotic margins[58]. The injured chondrocytes and extracellular matrix are not undergoing the remodeling process of the viable chondrocyte aggregation[58].

Superficial chondral lesions (Outerbridge I and II) are characterized by the absence of inflammatory cells and those involved in the repair process[56]. Unlike the deep osteochondral lesions, necrotic margins next to the superficial lesions pose an intense mitotic and metabolic activity[56]. In the absence of mesenchymal stem cells migration from bone marrow, this process mostly mediated by chondrocytes, is of short-term and ends within several weeks, resulting in an immature cartilage tissue[56].

## 2.2    Traumatic lesions

The principle of the therapeutic procedures is that an osteochondral breach with access to a vascular supply possesses healing potential, while the lack of stem cells within the superficial chondral lesions makes the repair process less likely[59,60].

The most frequent traumatic cartilage lesions of the knee area due to traumatic patella dislocations and anterior cruciate ligament tears. Lesions can vary in area and depth, even reaching the subchondral bone. The incidence of these are approximately 40% in patella dislocations and 16-40% in anterior cruciate ligament tears[61-63].

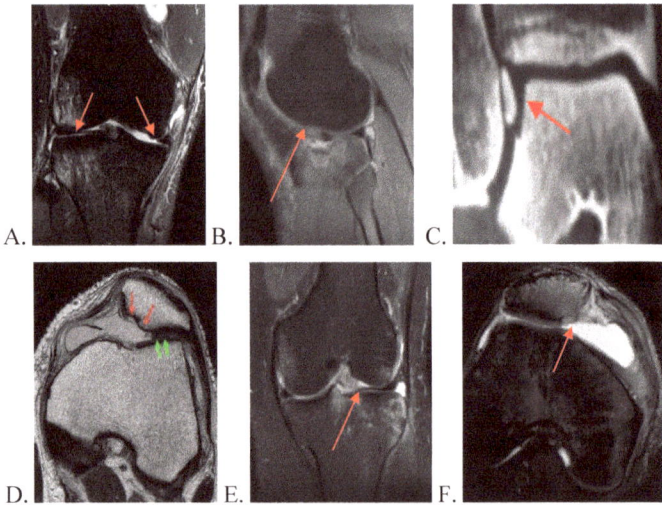

***Fig. 2.1.*** *Articular cartilage mechanism of injury.*

A.  ACL lesions
B.  Depression fracture of tibial plateau
C.  Osteochondral fracture of the external region of the talus dome
D.  Patella dislocation
E.  Tibial spine avulsion with osteochondral fragment
F.  MPFL avulsion with a cartilage fragment detached from the medial patella.

According to O'Donoghue, there are three mechanisms of injury for chondral and osteochondral fracture:

- *impact of bone ends* - by direct mechanism, such as patella compression on the femoral condyles, or during abnormal articular motion (sprains, dislocations), for example femoral condyles and posterior part of the tibial plateau osteochondral compression fractures following anterior cruciate ligament (ACL) (Fig. 2.1.A), depression fractures of the tibial plateau (Fig. 2.1.B) or osteochondral fractures of the talar dome (Fig. 2.1.C)

- *shearing* - for example by lateral translation of the medial patella over the lateral wall of the femoral trochlea, especially in children (Fig. 2.1.D)

- *avulsion fractures* - where parts of bone where ligaments and tendons attach, for example, tibial spine avulsion-fracture, can extend to adjacent cartilage (Fig. 2.1.E). Another example is the desinsertion of the quadriceps tendon or medial patella-femoral ligament from the patella, with the avulsion of the osteochondral fragment (Fig. 2.1.F)[64-66].

## 2.3    Ischemic lesions

According to data published by Edmunds and Shea, osteochondritis dissecans (OCD) is the focal destruction of subchondral bone which leads to instability and destruction of the overlying cartilage, leading to early arthritic changes within the articulation[67]. Following this process, a bone or osteochondral fragment withstands avascular changes, with or without the delamination of the overlying cartilage[68]. The etiology of this pathology is debatable, numerous causes were described such as trauma, genetic factors, ossification defects, vascular deficiency, endocrine pathologies, and malformations[68-74].

There are multiple classifications of this pathology. One of these, important for therapeutic management, divides the OCD according to the age of the onset, adult OCD and juvenile OCD. The juvenile OCD occurs to patients with active growth plates and has a better prognosis compared to adult OCD, with a healing rate of over 50% following conservative treatment[74-75].

Adult OCD is classified according to ICRS (ICRS – International Cartilage Repair Society) as the following types (Figure 2.2)[76]. Recently, the primary necrosis is considered to occur following a subchondral stress fracture on the background of osteopenia SIFK (Subchondral Insufficiency Fracture of the Knee), the secondary edema and rise of intramedullary pressure determining the onset of ischemia and necrosis[77-80].

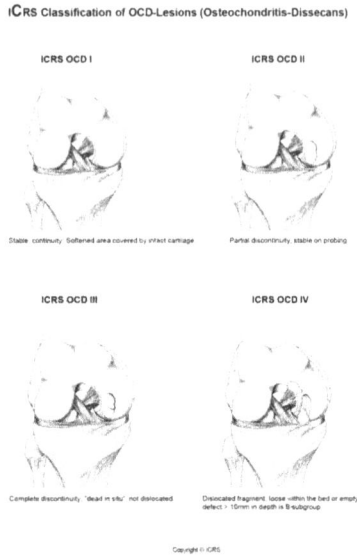

***Fig. 2.2.*** *ICRS classification of OCD lesions (Image from International Cartilage Repair Society (ICRS). ICRS Clinical Cartilage Injury Evaluation System). Available at: http://cartilage.org/society/ publications/icrs-score/. Accessed on February 26, 2017)*[76]

## 2.4    Factors associated with cartilage lesions

Static and dynamic changes within the articulation need thorough management of both the integrity of the structures, as well as the interaction of stabilizing elements.

Articular misalignment, due to abnormal distribution of loading forces, represents a predisposing factor for articular cartilage osteochondral damage, especially in case of a preexisting chondral lesion, when the forces focus on the margins of the defect, leading to its progression[81-84]. Frontal plane deviation varum and valgum associated risk of four to five times higher for the onset of unicompartmental osteoarthritis[85]. Nevertheless, recently it is considered that a major impact is represented by varum deviations, valgum deviations being considered only a borderline risk factor[82-86]. No studies so far showed a limitation of arthritis progression if the normal alignment is restored[87].

**Meniscus** represents a critical element for the integrity of knee articulation. Together with collateral ligaments, these are secondary stabilizers for pre prevention of anterior translation and external rotation of the tibia. The medial meniscus is more predisposed to tears compared to lateral meniscus[88]. The more rigid attachment of the medial meniscus to the deep layer of the internal collateral ligament leads to increased tensions during its sliding between the femoral condyle and tibial plateau, compared to lateral meniscus which has a less rigid fixation[89].

**Articular instability** is a complex pathology that involves all the functional components of the knee and is due to ligamentous insufficiency caused wither post traumatically or degeneratively[90]. The lesions can affect intra and extra-articular structures and are rarely isolated, usually, the ligamentous lesions are associated with lesions of the meniscus, osteochondral bone or other surrounding structures[90,91].

Ligamentous instability leads to the change in articular kinematics and dynamics which results in an increase in mechanical stress and uneven distribution of pressure at the bone ends, resulting in progressive damage of cartilage surface where the pressures are higher and the appearance of degenerative, instable and ill-defined cartilage lesions[92,93].

In the case of cruciate ligament lesion, we can have primary traumatic chondral lesions ("bone-bruises"), with well-defined margins, but there is also the possibility of iatrogenic injuries. Cruciate ligament repair, without repair of the meniscal lesions, can lead to an early failure of the surgery[84,94,95]. An excessive tibial slope (posterior inclination of the tibial plateau greater than 10 degrees), can lead to anterior knee instability, similar to the one caused by ACL insufficiency, and excessive loading of the posterior surface[84,96].

The role of collateral ligaments, as well as complex posteromedial and postero-lateral stabilizing structures is not to be neglected. Their lesions lead to complex instability which causes articular cartilage lesions where the stress is applied[97].

The patella-femoral joint is exposed to the apparition of chondral lesions due to its particular conformation, which determines the concentration of high forces on a very limited surface (forces equivalent to four times the body weight[98]). Patella-femoral instability can be secondary, cause by an anatomical anomaly, such as patella-trochlear dysplasia, or post traumatically, following patella dislocations with the lesion of MPFL (medial patello-femoral ligament[84,99]). Patellar misalignment (patella alta, patella baja and external subluxation of the patella) lower the contact surface with the femoral trochlea, determining the exponential increase of local pressures with the progressive mechanical deterioration of the cartilage, despite its increased thickness in the patella region (4 mm)[84,99,100].

## 2.5    Iatrogenic injuries

These can be produced by multiple mechanisms:

- **traumatic lesions**: intraoperative, accidental lesion of the cartilage surface, either during arthroscopically or open surgeries. An incriminated cause is also the use of intraarticular radiofrequency cautery, that produces heat and therefore can indirectly damage the cartilage cells[101,102].

- **toxic lesions**: in vitro and in vivo studies suggested the possible cytotoxic effects on chondrocytes of different local anesthetics used during intraarticular injection, especially bupivacaine and lidocaine, which can lead to the death of chondrocytes and chondrolysis[103,104].

- **the local effect of long-term corticoid administration** is debatable. The physiopathological mechanism is under debate and proposes a complex and discordant interaction. Therefore, *in vitro* studies support the chondrogenic effect due to the imbalance between matrix metalloproteinases (MMPs) synthesis and inhibition, while in *vivo* studies show a beneficial effect[105].

- **postoperative complications**: rapid progressive chondrolysis, can also occur in case of young athlete patients, following minimally-invasive surgical interventions, for example an arthroscopy can lead to a generalized destruction of the articular cartilage at the level of the articulation within 1 to 3 months, usually involving the external compartment of the knee, without any traumatic lesions involved, but possible toxic and ischemic mechanisms[106,107].

## 2.6    Degenerative lesions – osteoarthritis physiopathology

Arthritis represents a degenerative pathology of the articulations involving an osteochondral deterioration and remodeling, with progressive erosion of the articular cartilage associated with simultaneous changes in the bone and soft tissues[108-110].

Early histological data shows that the first phase is represented by the fragmentation of the extracellular matrix, followed by collagen fiber deterioration[109]. Chondrocyte proliferation as a response to chondral degradation is followed by diminished anabolic and proliferative activities[109]. Although the gene expression regarding the collagen synthesis is present in the late phases of the disease, their transcription, proteoglycans synthesis, and chondrocyte capacity to regulate apoptosis are diminished (Table 2.1)[109,111].

*Table 2.1.* *The stages of articular cartilage development and progression of osteoarthritis (Buckwalter et al. 1997)*[109]

| Stages | Description |
|---|---|
| **I. Interruption or alteration of cartilage matrix** | Interruption or alteration of matrix macromolecular network associated with an increase in water quantity can cause a mechanical lesion with the deterioration of macromolecular matrix or alteration of chondrocyte metabolism. Initially, collagen type II concentration remains at the same level, but it can associate deterioration of collagen network and decrease of aggrecan and proteoglycan aggregates concentration |
| **II. Chondrocyte response to lesion** | When the chondrocytes identify an interruption or change in the matrix, their proliferation is activated with the increase of both matrix synthesis and degradation. Their response can restore the integrity of the tissue, can maintain the damaged cartilage or can produce cartilage hypertrophy. This level of activity can be sustained for years. |
| **III. Decrease of chondrocyte response** | Failure of chondrocytes response to maintain the tissue viability determines the cartilage degradation, associated or preceded by the chondrocyte decrease in activity. The causes of this decrease are not clear, they are partially due to mechanical degeneration of tissues with chondrocyte lesions and partially to a decreased of chondrocyte receptors for anabolic cytokines. |

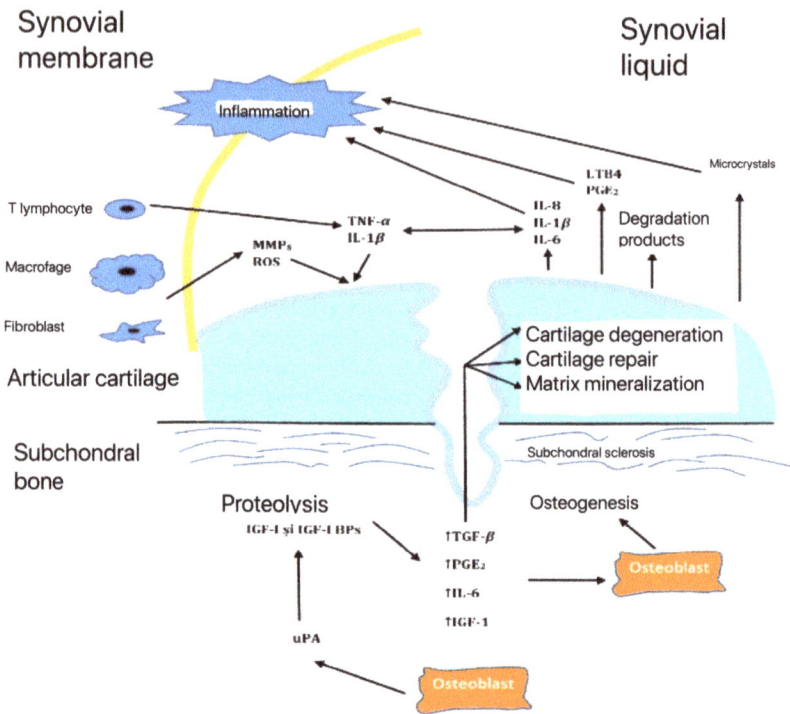

***Fig. 2.3.*** *Physiopathological pathways involved in osteoarthritis progression (Henrotin et al. 2005).[80]*

(MMP-matrix metalloproteinases; ROS-reactive oxygen species; TNF-a-tumor necrosis factor alpha; IL-1β-interleukin-1 beta; IL-6-interleukin-6; IL-8-interleukin-8; PGE2-prostaglandin E2; LTB4-leukotriene B4; uPA-urokinase-type plasminogen activator; TGF-β-transforming growth factor; IGF-1-Insulin-like growth factor 1; IGF-1 BPs-Insulin-like growth factor 1 binding protein).

## 3.    Diagnosis of cartilage lesions

Suspicion of cartilage lesions diagnosis is based on symptoms and clinical examination of the patient, but both clinical and paraclinical examinations need to be correlated.

### 3.1    Clinic diagnosis elements

They consist of symptoms and clinical examination of patients and can be localized in the shoulder, elbow, wrist, knee, hip, and ankle. Patient history can indicate the previous traumatism of the involved limb, or surgical intervention, which can indicate the presence of an iatrogenic injury.

From the symptoms point of view, regarding the shoulder, the patient accuses pain during passive and active mobilization, as well as inflammatory events. In some situations, chondral lesions are frequently associated with other types of pathologies such as rotator cuff tears, which influences the clinical examinations. Clinical tests are not specific for cartilage lesions, being similar in other types of pathologies such as impingement syndrome[112,113].

At the elbow, the chondral lesions produce pain during mobilization, local inflammation, and tenderness (e.g. pain during palpation of the lateral epicondyle in Panner disease)[114]. In the case of the wrist joint, the symptoms are pain and loss of range of motion[115]. In the absence of treatment and of healing, the cartilage lesions can extend leading to osteoarthritis of the whole joint, when crepitation, chronic pain, loss of range of motion and functional impairment occurs.

In the case of the cartilage lesions localized at the level of the inferior limb (knee, hip, and ankle), the patient accuses a deep pain made worse with physical activity, episodes of inflammation, as well as articular blocking. Nocturnal pain is present in most cases. The previously described symptoms impair the function of the limb, therefore subjective and objective scoring systems were invented to evaluate the impact of the articular pathology on the quality of life. Such a scoring system is IKDC 2000, which involves both subjective and objective criteria.

In the case of patellar chondral lesions, pain is localized in the anterior part of the knee, and it is exacerbated with activities such as going down or climbing stairs, squats. According to the degree of the cartilage lesions, they can further evolve to osteoarthritis, which is represented by chronic pain during physical activity or stance on the affected limb, relative functional impairment, bone crepitus, inflammatory episodes, and limb axis misalignment.

## 3.2 Paraclinic investigations

They are important both for diagnostic and treatment. The most frequently used investigation for the diagnostic are x-rays, but most information is obtained by magnetic resonance imaging (MRI).

### 3.2.1 Radiology

Radiological investigations are frequently made in the case of patients having the previously described symptoms. These investigations do not offer additional information in case of the early stages of the disease.

Radiologically, chondral lesions are represented as defects of the articular surface (e.g. in case of osteochondritis dissecans at the level of the knee and talus), or as bony areas surrounded by a radiolucent rim (when the osteochondral fragment is detached). This pathology can also determine the radiolucency of the entire area involved. Loose intraarticular bodies in the case of osteochondral lesion separation can also be observed. Indirect signs as intraarticular swelling or misalignment (e.g. patello-femoral syndrome) can also be observed. In the late stages of the disease, the localized or overall osteoarthritis is found.

### 3.2.2 Ultrasound

Conventional ultrasound does not provide valuable information due to the limited access within the joint. Recently, a minimally invasive high frequency (>10 MHz) ultrasound technique was developed to be used during arthroscopies and offers data regarding the cartilage lesions (e.g. biochemical and structural characterization of lesions)[116]. Nevertheless, the studies did not find it to provide more accurate information compared to MRI.

### 3.2.3 Computed tomography

As in the case of x-rays, the computed tomography can only detect cartilage lesions in later stages, its sensibility in the case of early stages is non-satisfactory[117]. Although magnetic resonance imaging is considered the best investigation for osteochondral lesions, a higher accuracy has been observed in the case of talus[118]. Therefore, this investigation can be taken into account in the case of MRI contraindications.

### 3.2.4 Computed tomography and magnetic resonance imaging arthrography

These two methods are carried out by injecting contrast fluid within the articulation. MRI arthrography has a sensibility of 41-79% and a specificity of 77-100% to detect chondral lesions[119]. Nevertheless, CT arthrography has better results in detecting osteochondral lesions compared to native MRI, and until recently it was considered the gold-standard in the diagnostic of cartilage lesions.

### 3.2.5 Magnetic resonance imaging

This type of examination is in most cases the final diagnostic investigation. Due to its reduced thickness and complex three-dimensional structure, the cartilage is best visualized with the help of magnetic resonance imaging devices of more than 1.5 Tesla[120,121]. Regarding the sequences, the T1 sequence is used for subchondral bone, while the T2 sequence is used for subchondral bone, as well as the interface between the cartilage and synovial fluid[122].

The most recommended MRI sequence to visualize the cartilage is 2D fast spin-echo (FSE) T2 weighted. The disadvantage is the relatively higher thickness slices, which can lead to skipping smaller dimension lesions[119].

Most cartilage lesions present a high signal in T2 (70%), while a normal or low signal is found in the other situations. Irregularities of cartilage, subchondral bone edema, synovitis, and loose intraarticular bodies indicate the possibility of cartilage lesions[122]. Studies point out a sensibility of approximately 87% and a specificity of 94% in the case of this sequence regarding the identification of osteochondral lesions in the knee joint[119]. MRI grading is found in Table 3.1. Nevertheless, the lesions measured during knee arthroscopies are 65% bigger compared to the ones measured during MRI examination[123].

*Table 3.1. MRI grading of osteochondral lesion[124]*

| Stage | Pathology |
|-------|-----------|
| 1 | Injury limited to the articular cartilage |
| 2a | Cartilage lesion with subchondral lesions and bone edema |
| 2b | Cartilage lesion with subchondral lesions without bone edema |
| 3 | Detached and undisplaced fragment |
| 4 | Detached and displaced fragment |
| 5 | Subchondral cyst formation |

Newer MRI techniques developed for a better examination of cartilage lesions. Therefore, T2 mapping is useful to visualize early lesions by determining the collagen and water content, as well as the collagen fibers orientation (Fig. 3.1)[121]. T2 mapping proved efficient in the characterization of tissue regenerations following surgical treatment[125]. Other biochemical methods of cartilage examination during MRI include delayed gadolinium-enhanced (dGEMRIC - quantifies the glycosaminoglycans at the cartilage level) and T1 rho (quantifies the proteoglycans at the level of the cartilage)[119,122].

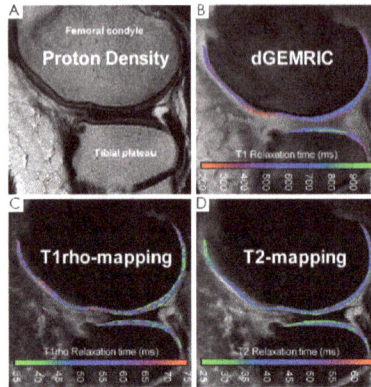

**Fig. 3.1.** *Quantitative MRI techniques. Conventional MRI does not show morphological cartilage defects (A). On the other hand, quantitative MRI techniques point out early osteoarthritic changes (B-D)[126]*

### 3.2.6   Biological markers

Literature tested numerous serological and intra articular markers to facilitate the diagnosis of cartilage lesions when other diagnostic examinations are not conclusive. Proteolytic enzymes as matrix-metalloproteinase MMP – 1, 3, 9, 13 and aggrecanase such as disintegrin, ADAMTS-4, and ADAMTS-5 in the articular fluid are under reserach[127]. Other markers of cartilage degradation such as CTX-II, COMP, and CS846 are also under investigation[128]. Currently, none of these markers are included in clinical practice.

### 3.2.7   Arthroscopy

Although diagnostic arthroscopy is no longer accepted because of the large number of non-invasive examinations previously presented, we present it because it determines the final diagnosis of osteochondral lesions. The most used classification of these lesions is Outerbridge, where these are classified in 4 stages (Table 3.2)[129].

***Table 3.2.*** *Outerbridge classification of knee osteochondral lesion.*

| Stage | Pathology |
|---|---|
| I | Softening and swelling |
| II | Superficial fissures of the articular cartilage on a surface not exceeding 1.5 cm |
| III | Subchondral fissures of the articular cartilage or a superficial fissure on a surface of more than 1.5 cm |
| IV | Exposed subchondral bone |

To study the articular cartilage regeneration or repair, an objective histological score was needed, such as O'Driscoll score[130]. Studies showed a weak correlation between the macroscopic scores such as ICRS and histological score[131]. Another score that analyses both macroscopically and microscopically cartilage lesions is the Wayne score[132].

## 4.    Treatment of cartilage lesions

### 4.1    Conservative treatment

This therapeutic option has the purpose to keep under control the symptomatology. It comprises the medicamentous treatment, weight loss, physical effort avoidance, rehabilitation procedures, physical therapy, orthoses, hyaluronic acid intraarticular injections, intake of chondroprotectors (like glucosamine, chondroitin sulphate, omega 3 fat acids, antioxidants), etc.[133,134]. The conservative treatment is addressed especially to the patients with small lesions, asymptomatic or incidentally discovered[133]. It is also recommended to those patients in which surgical treatment is expected to have poor outcomes, like smokers, obese or with intangible expectations[121].

In what concerns analgesic treatment, it is recommended to start with paracetamol or some non-steroidal anti-inflammatory drug. If symptoms persist, it is recommended to utilize light opioids like tramadolum. Locally applied capsaicin is another proven method for knee pain. Other adjuvant treatment like tricyclic antidepressants and anticonvulsive medication can also be used[135].

Physical therapy and rehabilitation are proven to have a positive effect for cartilage lesions located especially at the elbow, knee or ankle. This type of treatment is efficient inclusively in lesions like osteochondritis dissecans. Muscle strengthening leads to diminution of joint micromovements and thus it slows the progression of chondral lesions[136].

Intraarticular injections of hyaluronic acid or platelet-rich-plasma (PRP) have positive effects on cartilage regeneration[137]. Studies have shown that PRP infiltrations may have a better chondral repair effect than hyaluronic acid[136]. There are possibilities to inject intraarticular concentrates of mesenchymal stem cells, that have the capacity to reduce articular cartilage defects by regeneration with hyaline type of cartilage[138]. Another type of infiltration used in clinical practice is with corticosteroids. They reduce the inflammation and subsequently the symptomatology, but administered in small doses (<2-3 mg/dose) they do not have a negative effect on the articular cartilage[139].

### 4.2    Surgical treatment

Chondral lesions can benefit of surgical techniques meant to stop further progression of lesions, to repair or to replace the affected zones. The indications of surgical treatment are: grade III or IV Outerbridge lesions and lack of response to conservative treatment[140]. The patients must be warned regarding the relative long period of healing and rehabilitation and the negative prognostic factors, like obesity, smoke, age of over 55 years, other associated articular conditions, including inflammatory arthritis[140].

### 4.2.1 Direct treatment

The objective of treatment of chondral lesions is to obtain a complete healing of cartilaginous defect with hyaline type of tissue, perfectly integrated in the surrounding sane cartilage, in the final purpose of restoring a mobile, stable and not painful joint. Although a multitude of direct surgical procedures were described, there is no "gold standard" therapy for cartilage restoration at the moment. The choice of the treatment method is guided by factors that concern the patient and also the lesion type[140].

### 4.2.1.1 Intraarticular lavage and chondral lesions debridement

Cartilaginous lesions are sometimes correlated with a intraarticular inflammation caused by a synovial membrane hypertrophy, called synovitis. This produces an increase of intraarticular pressure by over-production of synovial fluid, causing pain. As symptomatic treatment, it is possible to perform an arthroscopic synovectomy and articular lavage[141]. In the same time, articular lavage will remove loose bodies and catabolic enzymes that stimulate the expansion of chondral lesions[142]. In this way, studies showed an improvement of functional scores and lower rates of recurrence, compared to articular aspirated punction[141,143]. But it is worth insisting that this method is purely symptomatic, without having a major impact on the cartilage condition.

Articular debridement consists in eliminating the affected chondral tissue and regularization of articular surfaces (Figure 4.1). After the procedure, a mobilization of stem cells from the subchondral bone may appear, with the role in articular surface repair processes[142].

### 4.2.1.2 Radiofrequency ablation

This technique is part of the arthroscopic procedures, and it consists of the removal of destroyed tissue and smoothening of the articular surface[143]. The radiofrequency probe performs cartilage ablation by formation of an electric arc with high temperature at its end[143]. Although this technique may have adverse reactions, like iatrogenic lesions of surrounding cartilage and osteonecrosis caused by thermic effects, it showed a decreased progression rate of chondral lesions comparative to mechanic debridement[144]. A recent development of this method is represented by the coblation probes that perform tissue ablation with electric arc formed in a chemical reaction, without heat production.

### 4.2.1.3 Microfractures (BMS - Bone Marrow Stimulation)

It represents the most common arthroscopic technique for treating chondral lesions. After lesion debridement, the subchondral bone is pierced with special sharp tools, facilitating formation of a super-clot at the level of chondral lesion (Figure 4.1). This clot acts like a scaffold for a fibro-cartilaginous tissue formation with the help of mesenchymal stem cells that migrate from bone marrow[140,141,145,146]. This repair tissue has a lower resistance comparative to normal cartilage, containing a greater proportion of type I collagen fibers[1].

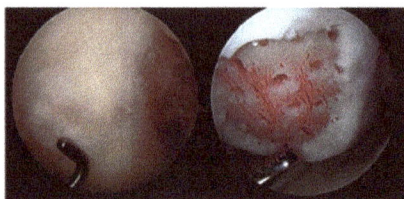

***Fig. 4.1.*** *Knee arthroscopy images. In the left picture it is observable a stage III chondral lesion on the medial femoral condyle. To the right, it is observed the migration of cells from subchondral bone after debridement and subchondral microfractures[147].*

Although there are controversies regarding the indications, this technique is most often utilized for young patients with less than 2 cm$^2$, well defined, type III or IV Outerbridge defects[140,145,147]. Studies have showed an improvement in overall knee function, but the effect may decrease after 18 months[146]. In comparison to autologous chondrocyte implantation, the results are similar for less than 2 cm$^2$ lesions[145]. The disadvantage of the method is represented by the inconsistent results at 5 years follow-up[148].

### 4.2.1.4 Osteochondral Autograft Transfer

This pure arthroscopic or arthroscopic guided technique can be performed with a series of devices like the Osteochondral Autograft Transfer System (OATS, *Arthrex, Naples, Fl, USA*), mozaicplasty set (*Smith & Nephew*) and Consistent Osteochondral Repair System (CORS, *Mitek*). The procedure consists in harvesting an osteochondral graft from non-weight-bearing areas of the knee and its transfer in the weight-bearing area with the chondral defect (Figure 4.2). The most often used donor zones are the lateral and medial part of the trochlea, the notch and terminal sulcus of the lateral femoral condyle[121]. Initially, the cartilaginous lesion is debrided until stable cartilage and then measured. The graft of cartilage and subchondral bone with a depth of 10-15 mm is then harvested. The disadvantages of this technique are given by the possible pathology of donor graft harvesting site and limited surface for harvesting[145].

The indications for osteochondral autograft transfer are femoral condyle lesions with less than 2 cm$^2$ surface, but satisfying results were also obtained in case of lesions between 2 and 4 cm$^2$ [145,149]. Marcacci reported good outcomes at 7 years follow-up for 77% of patients with femoral cartilage lesions treated with this technique[150]. In the same time, this technique offers better results than microfractures at the MRI control[140].

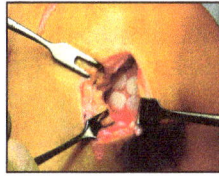

***Fig. 4.2.*** *Arthroscopic guided knee mozaicplasty with osteochondral autografts cylinders implanted at the chondral lesion site after debridement[149].*

### 4.2.1.5 Osteochondral allograft transplantation

This technique utilizes grafts harvested from individuals of same species[151]. Its advantages include a shorter operating time, smaller incisions and a no morbidity concerning donor site[141]. The immunosuppression is not necessary due to absence of cartilage vascularization[145,146]. The disadvantages are represented by a lower integration rate comparative to autografts, a more demanding technique, decreased availability and increased costs of allografts[140]. The indications are considered to be chondral lesions of more than 2 cm$^2$, revision of other cartilage repair procedures, osteochondritis dissecans of stage III and IV, osteonecrosis and posttraumatic reconstruction[146]. In case of posttraumatic defects, a survival of 85% at 10 years was obtained, while in case of osteochondritis dissecans a survival of 80% at 6 years was observed[152].

### 4.2.1.6 Xenografts

They are represented by grafts harvested from other species different than the recipient[141]. The most studied xenografts are obtained from pigs. But their use is limited due to possible inflammatory reactions[153]. In the same time, comparative to allografts, xenografts require longer integration period, thus having limited utilization[154].

### 4.2.1.7 Loco-regional transfer of pedicled osteochondral graft

The literature describes osteochondral graft transfers together with the vascular pedicle, as those for avascular necrosis of carpal scaphoid[155].

### 4.2.1.8 Osteochondral biphasic prosthetic implants

They are represented by hydroxyapatite cylinders covered with a layer of collagen or hyaluronic acid, that are implanted at the site of osteochondral defect in fitted size placements. These implants represent in fact bioresorbable scaffolds for cell ingrowth. They also have a mechanical and biological role for attracting stem cells, which will generate chondrocyte and osteoblast lines, parallel to implant degradation and bone regeneration[156].

### 4.2.1.9 Prosthetic arthroplasty

Arthroplasties are surgical interventions in which a joint is replace with an artificial implant. Arthroplasties can be total, in which the entire joint is replaced, or partial, in which only a part of the joint is replaced. Partial prostheses, for example in case of the knee, can replace a whole compartment (uni-compartmental prostheses) or just a portion of a compartment (e.g. UniCAP®)[157]. Regarding the localization, the most frequently performed arthroplasties are hip, knee and shoulder.

They are indicated in patients with severe and extended chondral lesions, but also in aged persons with limited regeneration potential. In case of young patients this type of surgical intervention must be reserved to cases in which the other techniques did not succeed or did not have indications. Another important aspect that recommends arthroplasties in elderly is represented by an earlier mobilization on the affected limb in comparison to the techniques of chondral regeneration or repair, which require protected mobilization[158].

### 4.2.2 Notions of tissue engineering

The failure of classic treatment methods led to the research of biological solutions in the attempt of healing the lesions. It is more and more emphasized the necessity of restoring cartilaginous cell stock, and not only that of fundamental substance. It is given credit for the potential of stem cells to differentiate towards cartilaginous line, but also to restore the subchondral bone, essential element for support and nutrition of cartilage. There are attempts to directly use the cartilaginous cells for local proliferation at the site of the defect, under the influence of biological stimuli, without the need for laboratory manipulation[159].

It is also more and more attempted to develope implants with a biological role in restoring subchondral bone, supporting scaffolds to which it is possible to add cellular elements that under the influence of biological factors, cytokines and specific growth factors, are redirected towards restoring the cartilaginous and bone cells. In this medical field tissue engineering of biocomposite materials found a proper place to extensively develop[160-162].

### 4.2.2.1 ACI - autologous chondrocyte implantation

It consists of harvesting chondrocytes from the non-weightbearing zones, their cultivation, differentiation and multiplications in an *in vitro* environment. About 200-300 mg of tissue are harvested[140]. After laboratory processing, in a surgical intervention, the chondrocytes are implanted under a periosteal membrane sutured to the surrounding normal cartilage (Figure 4.3)[141,146]. The periosteal membrane is harvested from the proximal tibia[145].

The advantage of this technique is the formation of a hyaline cartilage instead a fibrocartilage as it happens in other types of interventions[140]. The disadvantages are represented by the necessity of two interventions separated by a long time interval (about six weeks), the high costs, the impossibility of all arthroscopic procedures and the possible hypertrophy of the periosteal membrane, rendering this technique to be utilized only as an alternative to other direct treatments[140,145]. The indications are grade III and IV Outerbridge defects with larger sizes (2-10 $cm^2$) and revisions of other repair methods[145,146]. The specialty studies showed good results in about 82% of patients at 7 years after the intervention[146].

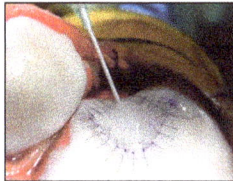

***Fig. 4.3.*** *Autologous chondrocyte implantation in the knee, covered by a periosteal membrane sutured to surrounding cartilage[135].*

*MACI - Matrix-induced Autologous Chondrocyte Implantation*

It appeared as an enhancement of autologous chondrocyte implantation, using an artificial collagen membrane (matrix) instead of the periosteum, not requiring its suture, and thus preventing the possible migration of chondrocytes in the articular space[163]. The results also showed better clinical scores for this technique[164]. The indications are represented by deep focal lesions between 2.5-4 $cm^2$ in patients with intense physical activity and more than 4 $cm^2$ in general population[165].

### 4.2.2.2 Multipotent mesenchymal stem cells

Stem cell have the capacity of dividing into two identical stem cells. They can differentiate in various types of cells, called pluripotent cells. These differentiate cells are the mesenchymal stem cells. At their turn, these can differentiate into multiple lineages: osteoblasts, chondrocytes, myocytes or adipocytes. The differentiation is controlled by growth factors, cytokines and other signaling pathways[166].

The use of stem cells for cartilage lesions treatment is a technique under development. It has the theoretical advantage of restoring hyaline tissue in the place of the defect, thus resulting a structure closer to the native cartilage, when compared to other techniques[167]. Various types of stem cells were studied for this purpose: human embryonic stem cells, pluripotent stem cells and multipotent stem cells – MSC. So far, none of the above mentioned was able to truly produce functional hyaline cartilage in the chondral defect in humans[168].

At this moment, the valid sources of stem cells are the following:

*Cells from bone marrow aspirate concentrate (BMC)*

Technically, the harvest is done by aspirating medullar puncture, usually at the level of the iliac crest, followed by centrifugation for obtaining a bone marrow concentrate (BMC). Optionally, a culture of MSCs can be done in order to test the proliferation capacity[169]. Their use at the defect site can be done in association with microfractures, by direct injection, but more efficiently they can be seeded on a scaffold of hyaluronic acid or collagen, in a custom fitted shape, during a single surgical intervention[170,171]. These stem cells can be activated with growth factors or cytokines. They are capable of restoring normal cartilaginous tissue and the method is cheaper and simpler than the ACI[172].

*Adipose derived stem cells (ASCs)*

At the level of adipose tissue there is a great concentration of stem cells that have the ability to develop in multiple directions, towards formation of cartilage, bone, muscle and adipose tissue, of course. Starting from this fact, the idea was promoted of harvesting and utilizing these cells for repair of chondral lesions, by a procedure similar to that of MSCs[173]. The advantages of this method would be: facile harvesting, great cells concentration that does not decrease with age, thus not necessitating additional cultivation and it is done in a single intervention (injection). The disadvantages are represented by a heterogenic cell mixture and increase formation of adipose tissue. But anyway, the experimental results are promising[174].

### 4.2.2.3 Scaffolds utilized in clinical practice

In cartilage tissue engineering, the scaffolds have the role to shape the hydrogels or porous matrices. Cartilage implanted 3D scaffolds are made of porous materials and apart from being a carrier for cells, growth factors or genes, they structurally reinforce the defects and prevent surrounding tissue to interfere with them[175].

The biomechanical properties of articular cartilage imply the capacity to preserve its shape and to restore the structure after loss of liquids. For scaffolds manufacture natural materials were used (collagen, gelatin, fibrin, chitosan, alginate, hyaluronan, silk), as well as synthetic biodegradable macromolecules (polylactic acid, polyglycolic acid, polycaprolactone). Collagen seems to be the most promising biomaterial for scaffolds, as it presents a natural adhesion to the surface of cells and contains biological information capable of directing cells activity. More than that, collagen is well-known for its excellent biocompatibility and a decreased antigenicity, while its degradation products are not harmful for organism[175].

For scaffold preparation, the collagen fibrils have to be functionalized in order to enhance their mechanical properties and delay their degradation. The mixtures made of collagen and amine groups containing polysaccharides and chitosan, which is another natural biodegradable polymer, were also used. A composite made of collagen with PLGA fibers was used in articular cartilage tissue engineering, with an adjustable height from 200 μm to 8 mm. Another composite biomaterial made of natural and synthetic polymers is poly ε-caprolactone-collagen type II-chondroitin sulphate (PCL-g-COL-g-CS) which was used as a biomimetic matrix that promoted spreading and growth of chondrocytes[175].

### 4.2.3    Corrective treatment of associated factors

### 4.2.3.1 Resolving ligamentous instability

Ligaments lesions can lead to subsequent articular instability, with a potential to create or aggravate chondral lesions. Anatomic ligamentous reconstruction is indicated immediately after rupture in order to reestablish normal articular kinematics and prevent development of chondropathies[84]. At the level of the knee, the most important and most encountered procedure is the reconstruction of the anterior cruciate ligament (ACL). It is indicated to be performed concomitant to cartilaginous or meniscal lesions repair, by methods that recreate the normal anatomy, in order to recreate the normal articular physiology, like all-inside method[176]. Cases of traumatic avulsions of ligament insertions can nowadays benefit of the same modern treatment, like reinsertions with special implants (e.g. TightRope)[66]. Femoral-patellar instabilities need surgical treatment by extensor mechanism realignment and reconstruction of affected ligaments (MPFL – Medial Patello-Femoral Ligament)[84].

### 4.2.3.2 Realignment osteotomies

They consist in a realignment of normal limb axis, which will lead to a decrease of pressure in the affected compartment of the knee. High tibial osteotomies can be performed in two ways: closing wedge osteotomies (in which a triangular shaped bone fragment is excised from proximal tibia) or opening wedge osteotomies (in which a bone graft is introduced at the level of an osteotomy traject[176]), as we can see in figure 4.4. They are indicated in cases of uni-compartmental lesions, age less than 60 years, knee flexion above 120°, body mass index less than 27.5 and WOMAC functional score above 45 points[177]. High tibial osteotomies favor cartilage regeneration at one year after surgical intervention[178,179]. Its disadvantages include non-union, inability of weight bearing on the affected limb for a long period (4-6 weeks) and possible increase of tibial slope[180].

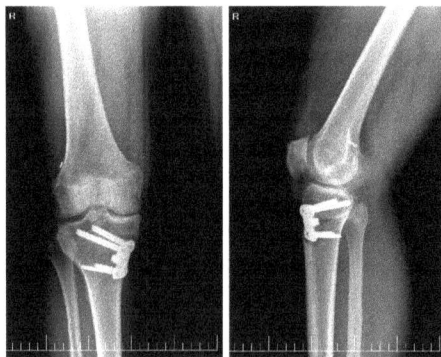

***Fig. 4.4.*** *High tibial osteotomy for medial compartment knee osteoarthritis in a 50 y.o. patient, concomitant with ACL reconstruction, in order to delay total knee replacement.*

### 4.2.3.3 Treatment of other protective factors

A special attention is given to the meniscus, due to its chondroprotective role by taking over the pressure off the cartilage surface. A meniscal lesion in contact with a cartilage lesion represents a contraindication for meniscectomy. It must be done with everything possible to preserve the meniscus, including the degenerative lesions, and in cases in

Materials Research Forum LLC
https://doi.org/10.21741/9781644900536

which this is impossible, it must be taken into consideration the possibility of meniscal transplantation[84].

Another protective factor is the integrity of subchondral bone. The articular fractures benefit nowadays of modern treatments like arthroscopic assisted reduction and minimally invasive percutaneous osteosynthesis. In these cases, the practice of arthroscopy brings a positive impact to the complete treatment of articular lesions, the key of therapeutic success[65].

## 5. Study 1. In vitro behavior of mesenchymal stem cells on different biomaterials

### 5.1 Introduction

Regenerative medicine is a growing field that makes an important contribution to medical science. Multipotent stem cells along with biomaterials are key elements of regenerative medicine, and the ability to differentiate and renew these cells is essential in restoring and repairing the tissues and organs of living organisms.

Hematogenic bone marrow is the main source of adult stem cells, but cells with similar phenotypic and functional characteristics have been identified in other tissues such as liver, lung, umbilical cord blood, kidney tissue, adipose tissue[181-183], exfoliated deciduous teeth, apical papilla root[184], palatal periosteum[185], periodontal ligament, neural crest between the Meisser corpuscles and the neurites of the Merkel perikaryons, components of hard palate[186,187] etc.

The capacity and characteristics of these cells are demonstrated in *in vitro* and *in vivo* studies. Under appropriate *in vivo* conditions, these cells have a potential for differentiation to specific cell lines, such as osteocytes, chondrocytes, adipocytes, myocytes, neurons, hepatocytes, beta-secreting insulin cells or keratocytes[188-191]. Based on the exposed features, these cells are of immense value for allogeneic, autologous and heterologous transplantation.

### 5.2 Work hypothesis

In this context, the main objective of our *in vitro* study focused on the isolation of mesenchymal stem cells from adipose tissue and hematogenic bone marrow, their characterization, the assessment of biocompatibility and the proliferation capacity of different biomaterials taken into study, including a newly developed collagen biomaterial, found in the early stages of *in vitro* and *in vivo* testing.

Through our experiments, we wanted to demonstrate that this new type of collagen scaffold is biocompatible, has no residual cytotoxicity and allows good adherence and proliferation of mesenchymal stem cells.

### 5.3 Materials and methods

The research was conducted between July 2015 and June 2016 within the Department of Breeding, Obstetrics and Veterinary Gynecology, in collaboration with the Department of Surgery USAMV Cluj-Napoca, Orthopedic-Traumatology Discipline within the University of Medicine and Pharmacy "Iuliu Hațieganu "Cluj-Napoca, Atomic Force

Microscopy Laboratory within the Physical Chemistry Center of the Babes-Bolyai University Cluj-Napoca and the Center of Electronic Microscopy within the Faculty of Biology, UBB Cluj-Napoca.

The study was conducted with the approval of the Ethics Committee of the University of Medicine and Pharmacy "Iuliu Haţieganu" Cluj-Napoca, no. approval 237/19.06.2014, and the experiments were conducted in compliance with the EU Directive no. 63/2010 and law no. 43/2014 in Romania. All treatments were carried out in view of animal welfare.

### 5.3.1  Cellular biological material

The biological material used was represented by the hematogenous bone marrow (n = 10) and adipose tissue (n = 6) harvested from healthy clinical sheep (n = 12), aged between 24 and 36 months (average 30 $\pm$ 6 months), average weight 50 kg ($\pm$ 5 kg). The harvesting of biological materials was carried out after general and local anesthesia was performed. The biological material was processed using the Concemo® kit (*Proteal Bioregenerative Solutions S.L., Spain*) to obtain the unicellular suspension.

Cellular suspensions have been added to the DMEM/F12 (*Sigma-Aldrich*) propagation medium supplemented by 10% fetal bovine serum (FCS, GIBCO), 1% antimycotic antibiotic (*Sigma-Aldrich*), 1% non-essential amino acids (*NEA, Sigma-Aldrich*), were evaluated immuno-phenotypical (*CD44 Hermes 1, Abcam*) and incubated at 37°C in a micro-climate enriched by 5% $CO_2$ and humidity 90%. The passage was performed at a concentration of 60-70%. For this purpose, the cultures were washed with PBS treated with Tripsin-EDTA solution (*Sigma-Aldrich*) for 10 min at 37°C, and the resulting cell suspension was centrifuged at 1200 rpm for 7 minutes. For subcultivation, a concentration of $5 \times 10^3$ cells/plate was used. The subcultures obtained were immunophenotypically characterized.

### 5.3.2  Tested biomaterials

Both collagenic biological materials useful for cartilage regeneration and anorganic calcium phosphate derivatives have been tested for use as bone substitutes, taking into account the possibility of combining them in biphasic scaffolds for use in repairing osteochondral defects (Table 5.1).

We tested for the first time a new type of collagen I + III (B3) implant obtained from the MatriBone Ortho product by decalcification procedures so that the anorganic phase of calcium salts was removed and a three-dimensional collagen I and III fiber network remained, which was further tested *in vivo* and implanted in ovine knee cartilage (see

Chapter 7). This derivat product is not available on the market and has been obtained by collaborating with the *Biom'Up* company, being delivered sterile in individual packaging, fulfilling all conditions for use *in vivo*.

***Table 5.1****. Tested biomaterials.*

| No. | Commercial denomination | Presenting form | Origin | Composition | Mechanism of action | Indications |
|---|---|---|---|---|---|---|
| B1 | Atlantik® *Medical Biomat SARL, France* | Granules | Synthetic | 70% Hydroxiapatite HAP + 30% Beta Tricalcium Phosphate (β-TCP) | Osteoconduction. Rapid resorption (β-TCP) + slow resorption (HAP) | Bone defects |
| B2 | Eurocer 400®, *FH Orthopedics SAS, France* | Granules | Synthetic | 55% Hydroxiapatite HAP + 45% Tricalcium Phosphate TCP | Osteoconduction. Rapid resorption (TCP) + slow resorption (HAP) | Bone defects |
| B3 | Matri Bone Ortho® demineralizat, *Biom'Up, France* (necomercializat) | 3D fibrilar netwok | Natural | Porcine Collagen Types I + III, eventually rests of Hydroxiapatite, Calcium Phosphate | Osteoconduction. Resorption and replacement with type II collagen (osseocartilaginous tissue) | Cartilage defects |
| B4 | Chondro-Gide® *Geistlich Pharma AG, Switzerland* | Bi-layered network | Natural | Collagen type I layer + Collagen type III layer | Resorption and replacement with type II collagen | Cartilage defects |
| B5 | Cova Max® *Biom'Up, France* | Mono-layered network | Natural | Type I acellular porcine Collagen | Resorption + guided regeneration | Bone defects |

### 5.3.3   Morphological characterization of the collagen matrix surface

Chondro-Gide bilayer (B4) is a collagen matrix designed for the regeneration of articular cartilage defects. For this purpose, the producer company had two surfaces of the collagen membrane: a rough one to ensure the fixation of the cells that are responsible for the regeneration of the cartilage, and the other surface being smooth and with a compact texture of collagen fibers to help *in vivo* fixation of the matrix and to prevent the migrations of the condrocytes from the desired workspace. This is an element of novelty that needs to be investigated in a morphological point of view.

The collagen implant resulting from the demineralization of Matri Bone (B3) is a complex composite material consisting of matrix collagen I/III, which may still contain traces of hydroxyapatite or tricalcium phosphate remaining after the demineralization process. In the composite form, the role of biomaterial is a resorbabyl osteo-conductive substitute. In the demineralized form the role should be the mold for the reconstruction of cartilaginous tissue. The destination of this biomaterial implies a specific morphology, compatible with that of cartilaginous tissue. Thus, by the atomic force microscopy, the surfaces of the two implants were investigated comparatively.

### 5.3.4 Evaluation of cell adhesion, proliferation and migration capacity of mesenchymal cells on demineralized MatriBone Ortho implants (B3)

To assess cell adhesion, the cultures were fixed with 4% paraformaldehyde and stained with DAPI (4,6-diamidino-2-phenylindole) and the nucleus of cells with TRITC (tetramethylrhodamine isothiocyanate). The cultures were evaluated with the Zeiss Axiovert D1 inverse phase microscope at a wavelength of 340/360 nm for DAPI and 546 nm for TRITC. Cells cultivated on plates without MatriBone Ortho membrane were used as a negative control.

For testing the functionality of mesenchymal stem cells obtained from sheep haematogenic bone (sheep line 2), in addition to test the adhesion capacity, it was chosen to test also the differentiation capacity on the MatriBone Ortho (B3) derived membrane. The cell suspension at a concentration of $3 \times 10^3$ was added to the pieces of membranes B3 (10 mm × 5 mm). The degree of adhesion, cell morphology and ability to form spheroids were evaluated after 2h, 24h and 3 days. Cellular proliferation was evaluated by Alamar Blue test, each tested in triplicate.

To assess the migration potential of ovine mesenchymal cells on B3 implants, cells at a concentration of $3 \times 10^3$ cells (passage 4) were aggregated in suspended drops in the propagation medium. After 48 hours, the formed aggregates were harvested and added to the culture plates provided with demineralized B3 membranes. After 24 hours, the migration potential was evaluated first by measuring the migration area. As a negative control, aggregates grown on culture plates were used. For calculation, the following formula was used: Migration Capacity = Migration Area - Aggregate Size. The evaluation was carried out in triplicate.

### 5.3.5    Evaluation of cellular adhesion and cytotoxicity on biomaterial

To evaluate the biocompatibility of the biomaterials studied, sheep mesenchymal stem cells (sMSCs – sheep Mesenchymal Stem Cells, p6x) were used. Cell viability, cytomorphology, cellular-biomaterial interactions, and cellular proliferation stimulating capacity have been evaluated.

For this purpose, $2x10^4$ cells were added to biomaterials (10 mm x 5 mm, and in the case of those in the form of granules an amount of 0.2 mg/well). The degree of adhesion and the proliferation capacity of the cells on the selected biomaterials were evaluated after 2h, 24 h and 3 days. Cellular proliferation was evaluated by the Alamar Blue test, and the adhesion after intranucleary staining with DAPI (4',6-Diamidino-2-phenylindole).

To assess the proliferation and adhesion, 150 µl of Alamar blue were added to each well. Plates were incubated for 1 h at 37° C, after which the cell culture medium was added to other multicompartimentated plates and the fluorescence intensity was evaluated with the BioTek Synergy 2 plate reader (540 nm excitation, 620 nm emission). After 3 days, the biomaterials (n = 5) cultivated with sMSCs were fixed with 2.7% glutaraldehyde (*Electron Microscopy Sciences, Hatfield, PA, USA*) in 0.1 M PBS. The implants were washed three times with PBS and scanned using the electronic scanning microscope (ESM).

### 5.4    Results
### 5.4.1    Characterization of stem cell populations

For the isolation of the mezenchimal stem cells of the bone marrow from the iliac crest, in addition to the filtering isolation method, centrifugation and differentiated extraction was chosen using the Concemo® kit (*Proteal Bioregenerative Solutions S.L., Spain*). After isolation, the cell suspension in the bone marrow concentrate (BMC - Bone Marrow Concentrate) was immunophenotized to detect CD44 positive cells (Figure 5.1).

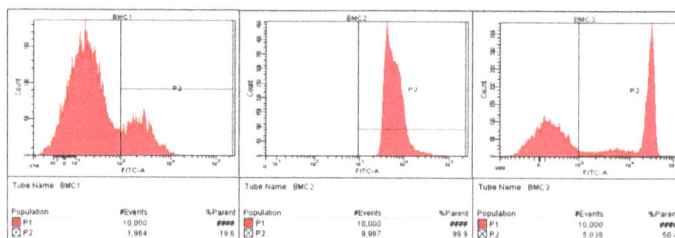

***Fig. 5.1.*** *Immunophenotypic analysis by FACS method of the isolated cells.*

The overall analysis of the results reveals a fairly wide variability between samples of BMC taken, registering an average of 40.88 ± 3.38 percent of stem cells in the samples taken (Figure 5.2).

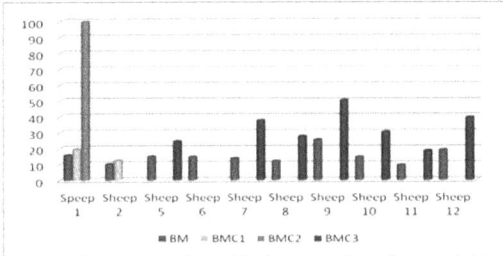

**Fig. 5.2.** *Global analysis of stem cells concentration in BMC.*

After the immunophenotypic analysis, cell suspensions isolated from 4 samples were cultured in a microclimate favorable to cell development and division. Examination of morphology and the degree of adherence of isolated cells was performed daily (Figure 5.3). The degree of adherence of the cells was examined in primary culture in the first 72 hours.

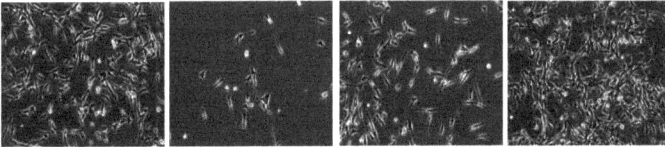

**Fig. 5.3.** *Aspect of cultures after 72 h.*

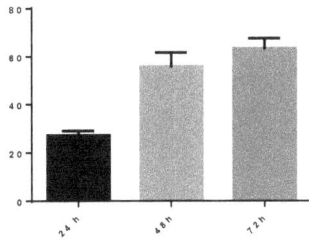

**Fig. 5.4.** *Global analysis of attachement degrees (0-80%).*

The overall analysis of the results shows statistically significant differences in the $p <$ 0.0001 (1-Way Anova test, table 5.2).

Twenty-four hours after isolation, the cells in the hematogenous bone marrow and the adipose tissue presented a marked cellular heterogeneity, with the predominance of round cells, fusiform, and in the case of adipose tissue of the polyhedral ones. The attachment rate was on average 29.33% ± 1.84 (SE). After 48 hours, the degree of attachment showed an average increase of 58.33% ± 5.72 (SE), and after 72 hours the degrees of attachment were on average 66.66% ± 4.24 (SE) (Figure 5.4).

***Table 5.2****. Results of statistical analysis regarding the attachement degree.*

| ANOVA summary | |
| --- | --- |
| F | 100.5 |
| P value | < 0.0001 |
| P value summary | **** |
| Are differences among means statistically significant? (P < 0.05) | Yes |
| R square | 0.9437 |
| ****very high significance p<0.0005 | |

The confluence of the cell cultures was recorded after day 15, when the first passage was made. After the first passage the cultures were homogeneous with the predominance of the cells with fibroblasts morphology and a sharp decrease of those with round morphology. The subculture cells, after a few days after the passage, began to organize in multiple colonies, these colonies being made of fusiform cells with accentuated bipolarity. In the following days, cell colonies gradually expanded and merged. The cells after the first passage were evaluated immunofenotypical, the results showing positivity for the mesenchimal CD44 marker, the global analysis indicating an average of 96.67 ± 3.20 (Figures 5.5 and 5.6).

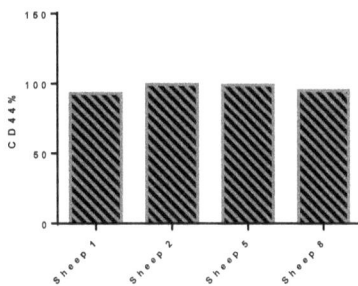

***Fig. 5.5****. Global analysis of attachement degrees.*

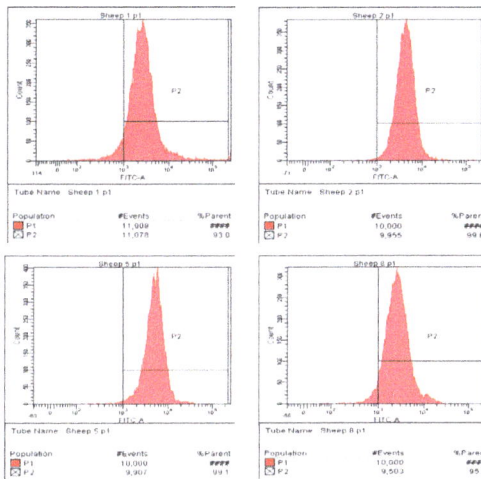

***Fig. 5.6.*** *Immunophenotypic analysis of cells from the 4 cultures, after first passage.*

### 5.4.2 Assessment of the degree of proliferation and mesenchymal cell migration capacity on demineralized MatriBone Ortho implants

Cell proliferation on B3 biomaterial was assessed by the Alamar Blue test. The analysis of the results indicates a significantly higher degree of attachment and proliferation compared to control culture (on the glass plate). The demineralized MatriBone Ortho membrane has a significantly higher proliferative effect compared to control culture, both in adhesion and proliferation, the differences being statistically significant $p < 0.05$ (Figure 5.7).

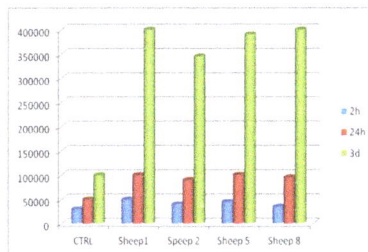

***Fig. 5.7.*** *Degree of attachement and cell proliferation on B3 – results of Alamar-Blue test.*

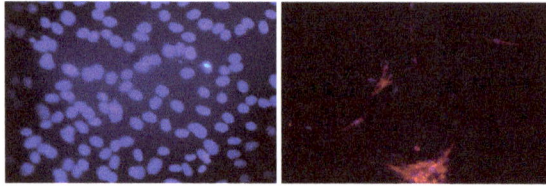

***Fig. 5.8.*** *Mesenchymal cells attached on demineralized MatriBone Ortho membranes,*
*coloured with DAPI, FITC, TRITC (original).*

For the assessment of the adhesion on biomaterial, the cells were fixed and stained with DAPI (Figure 5.8). The evaluation of the results was carried out using the fluorescence microscope. The attached cells were quantified in three randomly selected microscopic fields. After one hour, the degree of attachment to the control culture was on average $105\pm5.56$ (SE)/mm$^2$ compared to cells grown on demineralised MatriBone Ortho membranes, where the adherence rate averaged $199.5\pm1.29$ (SE)/mm$^2$. After 24 hours, the degree of attachment on this membrane was $199.75\pm3.59$ (SE)/mm$^2$, and after 3 days $201\pm6.0$ (SE)/mm$^2$ (Figure 5.9).

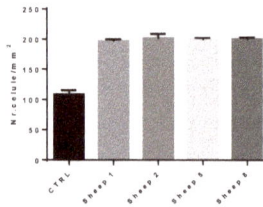

***Fig. 5.9.*** *Degree of attachement on the biomaterial of the 4 cellular lines tested (no. of*
*cells/mm$^2$).*

The overall analysis of the results shows statistically significant differences (Table 5.3). For the comparative assessment of the migration potential, the mesenchymal stem cells were aggregated and cultured in the control cultures and on the MatriBone Ortho demineralized biomaterial (B3). After 24 hours, the migration potential was assessed by measuring the migration area. The results were analyzed using the GraphPadPrism 6 statistical program.

***Table 5.3****. Comparative evaluation of cellular adhesion degree (substrate B3 vs. control).*

| F | 221.2 |
|---|---|
| **P value** | **< 0.0001** |
| P value summary | **** |
| **Are differences among means statistically significant? (P < 0.05)** | **Yes** |
| R square | 0.9888 |
| ****very high significance p<0.0005 | |

To assess the migration capacity on biomaterial B3, the mesenchymal stem cells in the 4 characterized lines were aggregated into the suspended drop. After 5 days, around the aggregates attached to the B3 membrane were observed the appearance of fusiform-like cells with accentuated bipolarity. As a control culture, the aggregates were added to biomaterial-free culture plates. For the biomaterial, the measured migration area was significantly higher compared to the control (Figure 5.10). The results are highly statistically significant (p <0.0001, Table 5.4).

***Table 5.4****. Results of statistical analysis – migration potential in control culture and on B3 biomaterial.*

| Column B | |
|---|---|
| **vs.** | **CTRL** |
| **Column A** | **vs.** |
| | **B3** |
| **Paired t test** | |
| **P value** | **< 0.0001** |
| P value summary | **** |
| **Significantly different? (P < 0.05)** | **Yes** |
| One- or two-tailed P value? | Two-tailed |
| t, df | t=29.57 df=3 |
| Number of pairs | 4 |
| **How big is the difference?** | |
| **Mean of differences** | **-0.1228** |
| **SD of differences** | 0.008302 |
| **SEM of differences** | 0.004151 |
| **95% confidence interval** | **-0.1360 to -0.1095** |
| R square | 0.9966 |
| **How effective was the pairing?** | |
| **Correlation coefficient (r)** | **-0.1656** |
| **P value (one tailed)** | 0.4172 |
| P value summary | ns |
| **Significant correlation? (P > 0.05)** | **Yes** |

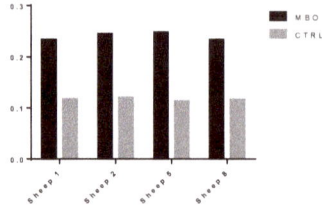

***Fig. 5.10.*** *Results of evaluation of migration capacity on B3 biomaterial (MBO) comparative with the control culture (CTRL).*

### 5.4.3   Morphological characterization of collagen membranes

The primary results of surface morphology obtained by AFM scanning of the Chondro-Gide matrix are shown in Figure 5.11.

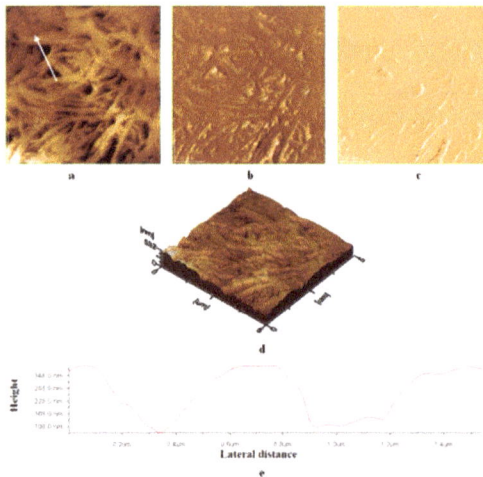

***Fig. 5.11.*** *AFM images of Chondro-Gide Bilayer:* ***a)*** *topographic image,* ***b)*** *phase image,* ***c)*** *amplitude image,* ***d)*** *3D representation of the topographic image,* ***e)*** *profile along the white arrow.*

*Scanned area 5x5 μm.*

The overall microstructural appearance of the Chondro-Gide membrane for the smooth part is shown in Figure 5.11a. A thick collagen braid is formed of horizontal fibers interlaced with vertically oriented fibers. The scanned surface has a maximum height of 532 nm and a roughness of only 59.3 nm, which correlates perfectly with the smooth side of the investigated foil. The height value is largely influenced by the presence of pores (the dark portions of Figure 5.11a) that connect with the depth layers of the foil. The collagen fiber network is also very well presented in the phase image, Figure 5.11b, where collagen has brown shades, and the empty spaces between the fibers appear in yellow, the depth of the pores appearing with the lightest yellow hue. The amplitude image, Figure 5.11c, shows that the surface of the membrane is free of packaging defects. Here, the contours of the pores that connect with the deep layers of the collagen film are distinctly visible.

The three dimensional representation of the topographic image, Figure 5.11d, highlights the way in which collagen fibers are arranged in the smooth surface of the Chondro-Gide membrane. We observe thicker fibers that form the basis of the structure, interposed with thinner fibers. The profile in Figure 5.11e highlights these thicker fibers with a diameter of around 400 nm, while the pores at their surface have an aperture with a diameter of about 600 nm. This network is definitely a porous one that can facilitate the circulation of biological fluids, but which successfully stops the flow of cells. The pore-specific depth is likely to locally increase the roughness value so that the profile has a value of 109 nm.

The thinner fibers in the smooth surface of the Chondro-Gide membrane are better observed at a scanning area of 1μm x 1μm, Figure 5.12. Surface topography captures a succession of several horizontally oriented fibers (Figure 5.12). The maximum image height is only 272 nm, with a roughness of 34.6 nm, indicating that the area of the thin fibers is even smoother on the one hand due to the lack of large pores and on the other hand due to the assembly of fibres being about the same level.

The macroscopic appearance of the demineralized MatriBone implant (B3) is highlighted with the help of SEM microscopy, as seen in Figure 5.13. At macroscopic level, it is observed that this implant presents a wrinkled look, very rough and similar appearance to a broken bony piece. The wrinkle appearance of the foil is explained by observing it at microscopic level, Figure 5.13, where the composite strips are folded into each other with widths between 20 and 50 μm and a thickness of less than 5 μm.

MatriBone-specific folds are likely to form a rather rough surface at the microscopic level, and right wrinkled at the macroscopic level. The situation is relatively unfavorable to the AFM investigation. However, it has been possible to position the scanning cantilever in some smoother areas of the surface where the folds show smoother surfaces with the side of at least 25 µm as seen in the upper left corner of the image in Figure 5.13b.

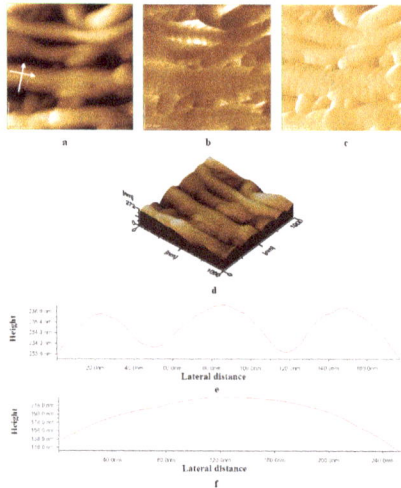

**Fig. 5.12.** *AFM images of Chondro-Gide Bilayer:* **a)** *topographic image,* **b)** *phase image,* **c)** *amplitude image,* **d)** *3D representation of the topographic image,* **e)** *profile along the white arrow.*

*Scanned area 1x1 µm.*

**Fig. 5.13.** *SEM images of demineralized MatriBone at various magnifications:*

**a)** *segment of 2 mm and* **b)** *segment of 100 µm.*

Therefore, the topography of the MatriBone surface with the AFM microscope is shown in Figure 5.14. At the microscopic level on the demineralized MatriBone implant, we found a smooth composite material fold over which a complex collagen fiber was formed, figure 5.14a. It has a diameter of 600 nm, as seen in the profile in figure 5.14e, and makes a complete loop within the image. Under fiber there is a composite material fold which at the bottom shows a very dense filament network, which is also observed in the phase and amplitude images, Figures 5.14b, c.

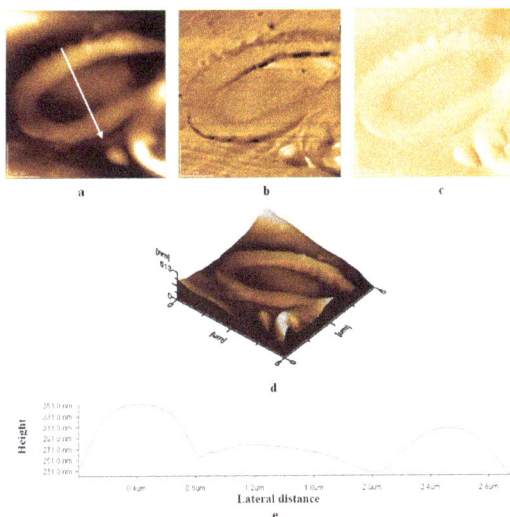

**Fig. 5.14.** *AFM images of demineralized MatriBone: **a)** topographic image, **b)** phase image, **c)** amplitude image, **d)** 3D representation of the topographic image, **e)** profile along the white arrow.*

*Scanned area 5x5 μm.*

In Figure 5.15, we were positioned by scanning directly on the mineralized collagen fiber of the MatriBone composite structure. Its topography has a peak height of 194 nm, resulting in a roughness of 53.6 nm. The appearance of the fiber is quite smooth, lacking in tropocollagen rings, but with some roughness on the surface, which correlates with the mineral particles inside the fiber. The phase and amplitude images, Figures 5.15b,c highlight a rather homogeneous structure, which correlates with the placement of the mineral material inside the colagenic material on the outside being only collage. The three-dimensional representation of the topographic image, Figure 5.15d, emphasizes the smooth appearance of the fiber surface. The profile drawn perpendicular to the fiber reveals a diameter of 600 nm, in accordance with the observations in Figure 5.14.

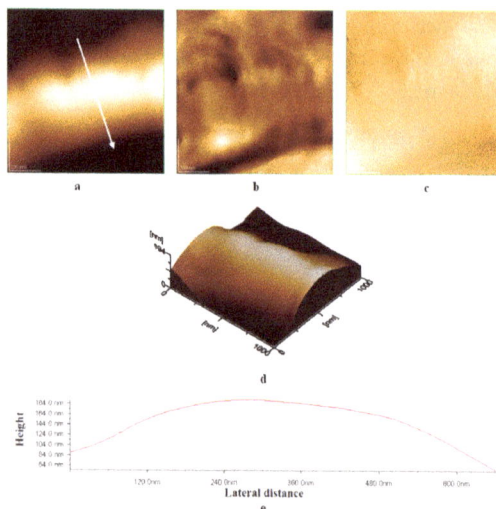

***Fig. 5.15.*** *AFM images of demineralized MatriBone:* ***a)*** *topographic image,* ***b)*** *phase image,* ***c)*** *amplitude image,* ***d)*** *3D representation of the topographic image,* ***e)*** *profile along the white arrow.*

*Scanned area 1x1 μm.*

### 5.4.4 Comparative microscopic evaluation on plate of cellular adhesion on biomaterials

The results of the comparative evaluation of the optical microscope of stem cell behaviour on different biomaterial will be presented below. Cellular adhesion was evaluated after nuclear staining with DAPI and is shown in Figure 5.16. It should be noted that the attachment was taken into account both on the biomaterial and on the cultivation surface.

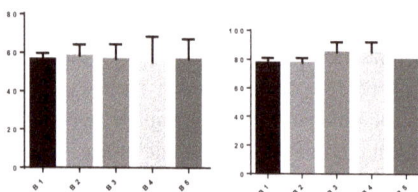

***Fig. 5.16.*** *Degree of cellular attachement on the biomaterials at 48 h and 3 days, respectively.*

### 5.4.5 Evaluation by electronic microscopy (SEM) of cellular adhesion and the implied cytotoxicity on biomaterial

At the biomaterial B2 (bone substitute) SEM analysis (Figure 5.17) reveals morphological changes in the cultivated cells, some of the cells have shown a fusiform morphology, others flattened, but most with round morphology. Also identified are cells arranged in conglomerate. The degree of attachment was high, the biomaterial favored the adherence of the cells. Cell aggregation and development of nodular formations may be associated with an early differentiation of the sMSCs cells towards the osteogenic line.

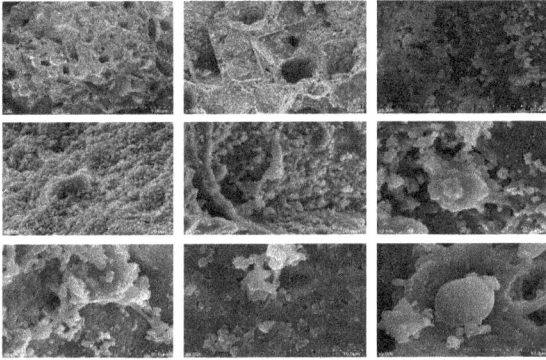

***Fig. 5.17.*** *SEM analysis of B2 biomaterial with cells (osteoconductive biomaterial made of HA+TCP granules).*

At B1 biomaterial (bone substitution HA + βTCP) the results were similar, after the addition of sMSCs and cell attachment, the fusiform cells with multiple filopodes and an increased percentage of round morphology cells were identified (Figure 5.18). The number of nodules was significantly higher compared to biomaterial B2. These nodules can be considered osteogenic nodules.

The degree of cell attachment for biomaterial B3 was significantly higher compared to the rest of the biomaterials, the cell colonies that were attached were organized as small groups or on the contrary, as multiple and fairly stretched clusters (Figure 5.19). The biomaterial is organized from several uniform filaments, the cells are attached to these filaments either on their own or in the form of interfibrillation bridges. The attached cells presented multiple phyllopetes and ondulopods (Figure 5.19). The cells were also identified among the porosities of the biomaterial. Osteogenic nodules and multiple cellular clusters and intercellular connections could be identified on the surface of the biomaterial. This material was proposed for *in vivo* testing on sheep.

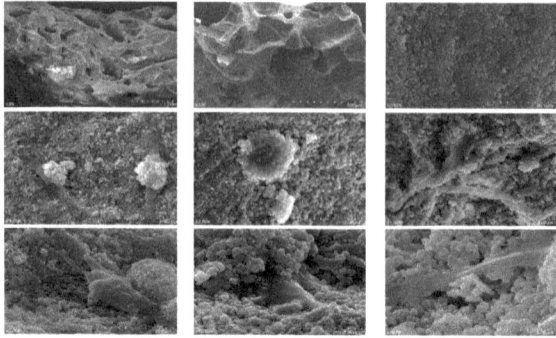

**Fig. 5.18.** *SEM analysis of B1 biomaterial with cells (osteoconductive biomaterial made of HA+βTCP granules).*

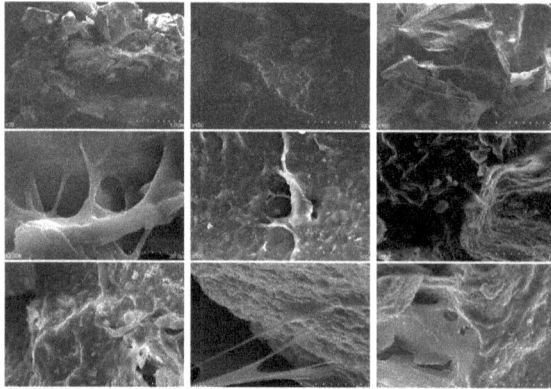

**Fig. 5.19.** *SEM analysis of B3 biomaterial with cells (3D collagen network).*

In B4 biomaterial (collagen I/III), the attached cells showed fibroplast-like morphology with numerous filopodes and ondulopods (Figure 5.20). The attached cells were interconnected and occupied the entire biomaterial surface 3 days after they were added. In B5 biomaterial (collagen I Covamax) the percentage of attached cells was lower compared to the other biomaterials tested. The MSCs cells retained the morphology of the fusiform cells and were also identified among the fine lamina of the biomaterial (Figure 5.21). No cellular clusters or osteogenic nodules have been identified.

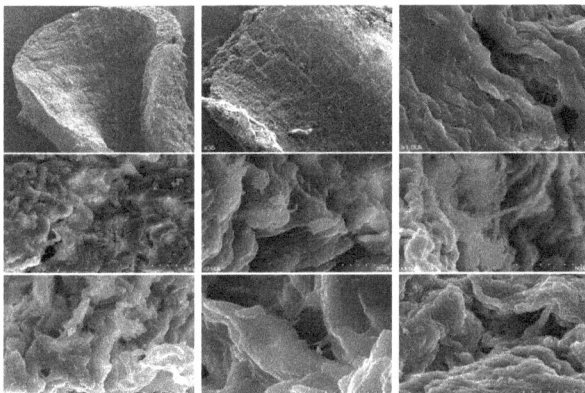

***Fig. 5.20.*** *SEM analysis of B4 biomaterial with cells (bi-layered collagen network).*

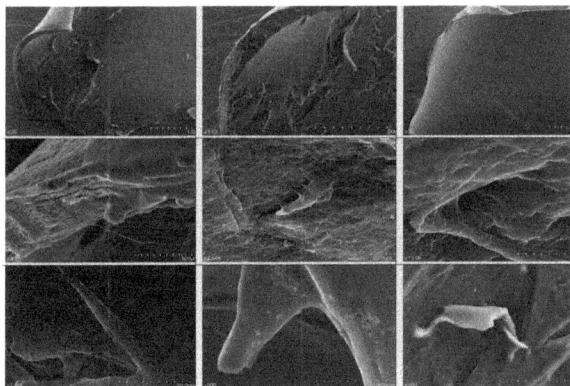

***Fig. 5.21.*** *SEM analysis of B5 biomaterial with cells (mono-layered collagen network).*

## 5.5    Discussions

With the help of the kits used, isolation and processing of the multipotent mesenchymal cells of the bone marrow from the iliac crest, from adipose tissue and from sheeps were accomplished following a protocol similar to those described in the literature[192-194]. Immunophenotic outcomes of CD44-positive cells (validated marker for multipotent mesenchymal cells[192,195]) indicate quite a wide variation between the harvested and analyzed samples.

Numerous scaffolds have been tested in recent years as a substrate for stem cell proliferation. Their *in vitro* testing implies the determination of the proliferation, migration and attachment capacity of cells[196,197]. In the case of innovative biomaterial B3, tested by us, we noticed a statistically significant superiority of it, compared to the control culture, in terms of the degree of attachment, migration and cellular proliferation.

The compactness of the collagen fibers arrangement is seen in the phase image (Figure 5.12b) in which the collagen fibers appear well attached to each other, so well and uniformly arranged that the tropocollagen rings in their composition with light stripes color can be seen. The tropocollagen rings are clearly outlined in the amplitude image, Figure 5.12c. The clear highlight of the fibre-specific tropocollagen rings is an additional proof of their purity and their assembly under native conditions.

The three-dimensional representation of the topographic image, Figure 5.12d, highlights the parallel horizontal alignment of thinner collagen fibers. Their diameter varies from 100 to 200 nm, as can be seen for example in the profile in Figure 5.12e. On the other hand, the profile of Figure 5.12e is drawn along the fiber and captures three successive tropocollagen rings, their diameters between 40 and 60 nm, which coincides with the observations in the literature for pure collagen[198]. The rough part of the Chondro-Gide membrane could not be investigated at the AFM due to the rugged surface exceeding 4 µm, which exceeds the specifications of the device used.

Returning to collagen fiber we notice that it has a matte surface without the specific striations of the tropocollagen rings, fact that is a pertinent indication in favour of internal mineralization of the fibre. Indeed, the nanostructured hydroxyapatite crystals and the remaining tricalcium phosphate have penetrated into the structure of collagen, causing the internal morphology of its assembly to be altered.

The fact that neither the surface of the fiber nor the surface of the fold shows any hydroxyapatite crystals means that they were very well incorporated into the collagen mass, which could not be extracted from the demineralisation process. Indeed, the nanostructured hydroxyapatite crystals and the remaining tricalcium phosphate have penetrated into the structure of collagen, causing the internal morphology of its assembly to be modified. The fact that neither the surface of the fiber nor the surface of the fold shows any hydroxyapatite crystals means that they were very well incorporated into the collagen mass, which could not be extracted from the demineralisation process.

From the topographic point of view, the maximum height of the demineralized Matri Bone implant corresponding to the scanned area in Figure 5.14 is 513 nm, resulting in a 59 nm roughness. We can conclude that at the microscopic level this implant is quite smooth.

The placement of mineral material inside the MatriBone collagen structures is likely to have a major interest in the interaction with the living bone, the direct interaction of the collagen from the outside with osteoblast cells may be a target for bone regeneration, and as it reaches the inward phase of the mineralization of the material, it to play the role of mineralization of new bone structures in course of formation.

The cells present at 3 days a comparable degree of attachment to the B3 implant (Chondro-Gide), which is currently a reference implant for the treatment of articular cartilage defects because it has proven its superiority in other similar experiments[199].

Electronic scanning microscopy is an appropriate tool for descriptive analysis of biomaterials[200], especially for the visualization of the interconnected fiber network[201], which is why we also used it in our study. The newly studied implant (B3) has all the characteristics of a good osteocartilaginous substituent, from an electronomicroscopic point of view: superior cell attachment, the organization in the form of groups or clusters, the presence of interconnected cell extensions and osteogenic nodules.

All these experimental results obtained with the decalcified MatriBone Ortho implanted (B3) encourage us to continue testing it *in vivo*.

## 5.6    Conclusions

Following our study on the isolation of bone marrow stem cells (MSCs) from the iliac crest (BMC on an ovine animal model) using the Concemo® kit and the evaluation of cell function on demineralized MatriBone Ortho biomaterial, we can formulate the following conclusions:

• hematogenic bone marrow and adipose tissue isolated cells can be easily processed using the Concemo® kit, but the immunophenotypic results indicate a fairly wide variability between the harvested and analyzed samples;

• until the first subcultivation, the cell cultures showed a marked heterogeneity, with the predominance of round and fusiform cells, with the confluence of culture on average after 14 days from isolation;

• after the first passage, positivity was recorded for the mezenchimal marker CD44, the overall analysis indicating an average of $96.67 \pm 3.20\%$;

• the degree of attachment and proliferation on demineralized MatriBone Ortho was significantly higher compared to control cell culture;

• the cells show a comparable degree of attachment on the B3 implant to the B4 implant (Chondro-Gide) at 3 days, which is currently a reference implant for the treatment of articular cartilage defects;

• the results of the assessment of the migration potential on the demineralized MatriBone Ortho indicate statistically significantly better results compared to the control culture;

In conclusion, we recommend the cultivation and transplantation of mesenchymal stem cells on the demineralized MatriBone Ortho implant, which is biocompatible, has no residual cytotoxicity and allows for good adherence and proliferation of mesenchymal stem cells.

## 6.    Study 2. Lipoaspirate fluid derived stem cells use for the treatment of cartilage defects. Pilot study on rabbit model

### 6.1    Introduction

The reconstruction of articular cartilage can be carried out using some specific methods of tissue engineering, namely by cell-seeded scaffolds. The three-dimensional scaffolds implanted in cartilage consist of porous materials, and apart from being a vehicle for cells, growth factors or genes, they structurally reinforce the defects and prevent the surrounding tissues from having access to their level[202]. Collagen appears to be the most promising biomaterial for scaffolds, since it has a natural adhesion to the surface of cells and contains biological information capable of directing cell activity[175].

Restoring the articular cartilage through hyaline tissue formation is still a challenge for surgeons and researchers. Mesenchymal stem cells (multipotential somatic precursor cells located in the perivascular areas of the connective stroma of adult tissues) seem to be an attractive perspective for cartilage repair due to their potential to express specific molecular markers, to continuously release growth factors and to differentiate on the chondrogenic lineage[203]. Most clinical and experimental approaches have so far been attempted to rely on the use of bone marrow stem cells (BMSCs) as a reasonable candidate for chondrocyte regeneration[204,205].

However, the low availability of the adult tissue source, the relatively invasive harvesting process as well as the low cell supply, make the bone marrow an inefficient and insufficient source of stem cells in this context. In recent years, stromal cells in the vascular-stromal compartment of adipose tissue have been proposed as a valid alternative because of their wider availability and comparable plastic properties[205-207].

The adipose tissue is indeed an excellent source of mesenchymal stem cells called ASCs (adipose-derived stem cells), with an average production of approximatv 5.000 colony-forming unit of fibroblasts CFU-F (fibroblast colony-forming unit) per gram of adipose tissue, versus approximately 100-1.000 CFU-F per milliliter of bone marrow[208]. Furthermore, adipose tissue is harvested through a minimally invasive procedure. However, most of the procedures based on these cells are time-consuming, technically difficult and require multiple interventions and *ex-vivo* handling, thus involving high costs, risk of contamination and local infections. The ideal procedure should be a single intervention, with minimal tissue manipulation[209].

Recently, the idea of obtaining stem cells directly from the liquid fraction of the adipose tissue aspirate LAF (lipoaspirate fluid) has been supported by relatively simple mechanical procedures, and these cells could at least theoretically be useful and valuable for cell-based therapies, just like as well as those obtained by laboratory processed lipoaspirate PLA (processed lipoaspirate)[210]. There is very little mention in the literature about the *in vivo* regenerative potential of cells from lipoaspirate fluid so far, and in particular, there are no data on the differences between the cells derived from the lipoaspirate fluid and the laboratory processed lipoaspirate regarding the initiation of the repair of articular cartilage.

## 6.2    Work hypothesis

The purpose of the study was to assess and compare the chondro-regenerative properties of cells from the lipoaspirate fluid and those from the laboratory processed lipoaspirate, on a preclinical model of cartilage defect in the knee joint of rabbits (Figure 6.1). The hypothesis of the study was that LAF would also benefit from trophic effects on cartilage regeneration, due to the presence of plastic stem cells alongside soluble molecules.

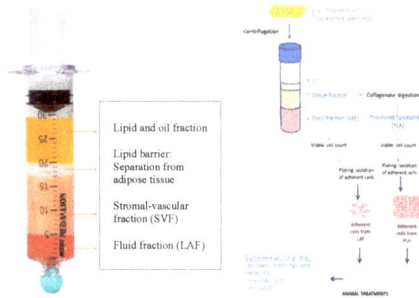

***Fig. 6.1.*** *Experimental protocole schematization.*

## 6.3    Materials and methods

### 6.3.1    Ethical statement

The present study was conceived as a pilot study. All experiments were conducted in accordance with the current practice guidelines and were approved by the Ethics Committee of the "Iuliu Haţieganu" University of Medicine and Pharmacy Cluj-Napoca, Romania, with the no. 340/05.06.2015 and in accordance with EU Directive 63/2010 and law no. 43/2014 from Romania.

### 6.3.2   Implants

For the surgical implantation procedure, commercially available collagen type I/III membranes Chondro-Gide® (*Geistlich Pharma AG*) were used. These membranes were impregnated with mesenchymal stem cells isolated from adipose tissue. For this procedure standard sterile instrumentary in aseptic surgical environment was used.

### 6.3.3   Cells isolation

Adipose tissue was prelevated from a healthy human donor (after signing informed consent specifying the use of biological samples for scientific purposes), divided into two tubes and alternately processed either using enzymatic treatment and centrifugation, or using the MyStem EVO® kit (*Bi-Medica Srl, Treviolo, Italy*) to separate the fat and the lipo-aspirated fluid[209]. The cells were counted, their viaiblity was checked and were stretched on the plates to isolate those adherent (CD105+).

Stem cells from adipose tissue obtained by standard digestion with colagenase (PLA - Processed Lipoaspirate) and lipoaspirate fluid (LAF cells - Lipoaspirate Fluid Cells) using the MyStem EVO kit, were cultured in Cole plots of 25 cm$^2$, in the incubator,  with 5% $CO_2$, humidity 90% and then expanded by 2-3 passages. The used culture medium was DMEM with 4.5 g/l glucose, 1% antibiotic, 2% mM L-glutamine and 10% FBS.

### 6.3.4   Seeding of cells on collagen membranes

To perform *in vivo* implantation, the cells were detached from the plastic plates by trypsinization: the culture medium was removed and the adhering cell layer was washed with sterile PBS. Two milliliters of 0.25% EDTA Trypsin solution was poured into the culture vessels and after 2-3 minutes, when the cells were detached, 5 milliliters of complete medium was added. The cell suspension was transferred to a 15 milliliter Falcon tube and centrifuged at 1200 rpm for 5 minutes. The cells were re-suspended in 1 ml of complete medium and counted with a Thoma count.

The sterile collagen membranes Chondro-Gide® (*Geistlich Pharma AG, Wolhusen, Switzerland)* with a diameter of 6 mm were arranged on Petri plates, then $1 \times 10^5$ LAF and PLA cells were added alternately to each plate. The seeding of cells on the membranes were visualized in contrast microscopy at 30 minutes and 24 hours (Figure 6.2).

### 6.3.5   Animals and their housing

Nine nine-month-old New Zealand white rabbits, average age 15 months (12-18 months), weighing 3.750±0.6872 kg, were used (mean ± 0.2173 DS). Prior to initiating the study, all animals were kept in quarantine and acclimated after separation from rabbit colony for

14 days. The animals were housed in standard cages with a total area of 4260 cm$^2$/cage, one animal per cage, with access to filtered water and granulated feed (*Granulated Combined Nutrition, Cantacuzino Institute, Romania*), ad libitum. The rabbits were cared for and kept in *Biobase* of the University of Medicine and Pharmacy Iuliu Haţieganu at a standard temperature of 20 ± 2°C, relative humidity of 45 ± 10 %, 12:12 h light/dark cycle (light on from 07 to 19 h).

### 6.3.6   Surgical technique

The animals were anesthetized using Diazepam (*Terapia SA, Romania*) at a dose of 1 mg/kg followed by Ketamine 35 mg/kg (*CP Ketamine, CP Pharma, Germany*) and Xylazine 5 mg/kg (*Xylazin Bioveta, Czech*). Prior to starting the surgical procedure, the anesthetized rabbits were immobilized on the operating table in lateral decubitus and the knee joint region was shaved. The preparation of the surgical field was done with alcoholic betadine solution (*Hexipharma, Romania*) and sterile adhesive drapes.

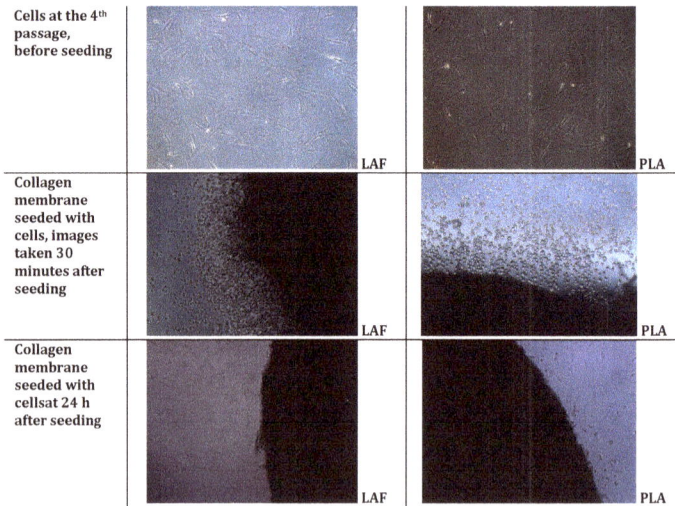

***Fig. 6.2.*** *Contrast microscopy images of cells seeded on membranes.*

Local subcutaneous infiltration with Bupivacaine (*Bupivacaine, Abbot Laboratories, USA*) was also performed. For the exposure of the joint, an internal parapatellar approach was practiced in the right knee. The osteochondral defect of 6 mm diameter and 3 mm deep was created with a biopsy punch and drill bit (Figure 6.3).

The animals were divided into three groups of 3 animals each, depending on the treatment applied: in **Group A** (control group) the defect was filled with non-seeded collagen matrix (Chondro-Gide alone), in **Group B** the matrix used was that seeded with LAF cells, and in **Group C** the matrix was seeded with PLA cells (Figure 6.4).

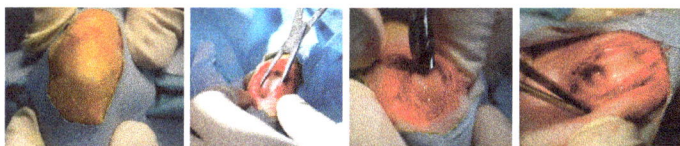

**Fig. 6.3.** *Incision and creation of the cylindric femoral trochlear defect.*

In all cases, the collagen matrix was fitted to cover the defect and glued in place with a tissue fibrin glue Tisseel Lyo® (*Baxter, Deerfield, IL, USA*). The surgical wound was closed in anatomical layers, 3-0 polyglycolic acid sutures (*Bicril Rapid 3/0, Biosintex*) and without drainage.

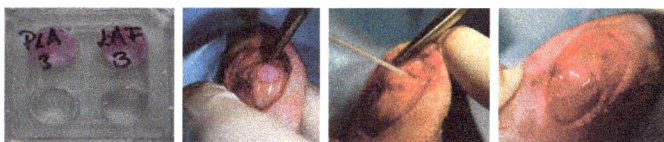

**Fig. Fig. 6.4.** *Filling of the defect and sealing the membrane on place.*

Following the surgical procedure, the animals received the following medication: Tramadol in 2 mg/kg (*Mabron, Medochemie Ltd, Cyprus*), daily for 7 days, Meloxicam 0.5 mg/kg (*Melovem, Dopharma, Romania*) for 3 days and Enrofloxacin 10 mg/kg (*Enrofloxacin, Pasteur Institute, Romania)* for 7 days. The wound has been disinfected daily for three weeks. Immediate mobilization was practiced within the tolerance limit. Health status and weight were monitored during recovery.

### 6.3.7 Evaluation of results

Animals were euthanized three months post-operatively, anesthetized according to the protocol described above, then administered potassium chloride i.v. The distal femur was dissected, removed and scanned by the micro-computer tomography immediately after harvesting. Using a mini-saw, two samples were taken from each rabbit. A sample was immersed immediately in liquid nitrogen for subsequent molecular analyzes and the other fixed in 4% paraformaldehyde for morphological studies.

The objectives of the study were: macroscopic and microscopic evaluation, quantification of gene expression responsible for cartilage proliferation and observation of internal morphology and microstructure. Immediately after sacrifice and dissection, a first evaluation was performed by two observers, according to the Wayne Cartilage Repair Score[211] (Table 6.1). This scoring system includes a 16-point scale based on four parameters: coverage of the defect with regenerative tissue, the color of the neocartilage, edges, smoothness and the color of the defect.

*Table 6.1. Wayne Scale for macroscopic evaluation of cartilage repair.*

| GROSS APPEARANCE | DEGREE | GROSS APPEARANCE | DEGREE |
|---|---|---|---|
| **Coverage** | | **Defect margins** | |
| >75% | 4 | Invisible | 4 |
| 50-75% | 3 | 25% circumference visible | 3 |
| 25-50% | 2 | 50% circumference visible | 2 |
| <25% | 1 | 75% circumference visible | 1 |
| No fill | 0 | Entire circumference visible | 0 |
| **Neocartilage color** | | **Surface** | |
| Normal | 4 | Smooth/level with normal | 4 |
| 25% yellow/brown | 3 | Smooth but raised | 3 |
| 50% yellow/brown | 2 | Irregular 25-50% | 2 |
| 75% yellow/brown | 1 | Irregular 50-75% | 1 |
| 100% yellow/brown | 0 | Irregular 75% | 0 |
| **TOTAL** | | | **16** |

After a 24 hour incubation with 4% paraformaldehyde buffer in PBS at 4°C and pH 7.5, the samples were washed in water for one hour and then decalcified for 15 days (*Decalcifier II, Leica*, according to the manufacturer's instructions, with daily change), dehydrated using the Thermo Scientific™ STP 120 Spin Tissue Processor (*Thermo Fisher Scientific, USA*) and sunken in paraffin using the Bio Optica Paraffin Dispenser DP500 (*Bio Optica, Milano, Italy*). The tissue samples were then cut into 15 µm thick sections with a Leica SM2000R microtome (*Leica Biosystems, Wetzlar, Germany*). Subsequently, the samples were stained with Alcian Blue for cartilaginous tissue[212].

In order to analyze the molecular differentiation potential in cartilage components, quantitative PCR (qPCR) was used to quantify the expression of the genes involved in cartilage differentiation. For this purpose, total RNA was isolated from the samples harvested from the three groups, placed in 6 wells plates using the RNeasy MiniElute Cleanup Kit (*Qiagen, Valencia, CA, USA*) according to the manufacturer's instructions, procedure described elsewhere[213].

An additional step of incubation on column was performed with the DNase enzyme to allow selective elimination of genomic DNA during the isolation process. Thus, reverse transcription in two steps and quantitative PCR were performed as previously described[214]. The resulting cDNA was used as template for qPCR to analyze expression of the genes encoding COL2A1 (Alpha 1 collagen type II), ACAN (Aggrecan) and SOX-9 (Y-box 9 sex determination region) involved in cartilage differentiation.

Method $2\text{-}\Delta CTt$[215] was applied to calculate the packaging differences in gene expression using the GAPDH gene (Glyceraldehyde 3-phosphate dehydrogenase) for normalization. PCR products were subjected to the melting curve analysis to exclude the synthesis of non-specific products. The oligonucleotide primers were designed using the Primer 3 software (http://bioinfo.ut.ee/primer3-0.4.0/); the primers sequences are reported in Table 6.2.

***Table 6.2.*** *Sequences of gene primers of interest*

| Gene | Direct Primer | Reveresed Primer |
|---|---|---|
| COL2A1_rabbit | AGAGACCTGAACTGGGCAGA | GAGGTCTGGCAGGAAGACAA |
| ACAN_rabbit | GGCCACTGTTACCGTCACTT | ATGCTGCTCAGGTGTGACTG |
| SOX9_rabbit | AGCTGAGTCCCAGCCACTA | GAGGTTGAAGGGGCTGTAGG |
| GAPDH_rabbit (housekeeping) | TGCGTTGCTGACAATCTT | ATTTGAAGGGCGGAGCCA |

Microstructure and internal morphology were obtained by microCT technique. Immediately after sampling, samples from groups B and C were scanned with a microCT (SkyScan 1172; *Brucker, Billerica, MA, USA*) at 80 kV and 100 mA using a 1 mm Cu + Al filter with step rotation of 0.4 and resolution of 20 μm. The calibration of the device was done with hydroxyapatite rods of 8 mm diameter (calibration phantoms). The samples were wrapped in paper, loaded in plastic Falcon tubes and moistened with saline solution, exactly like the phantom rods. The sections were rebuilt with the NRecon software (*Bruker, Belgium*) and analyzed using CTAn (*Bruker, Belgium*).

### 6.3.8   Data analysis

The continuous variables were analyzed to note the significant differences between groups. Bidirectional variance analysis (ANOVA) was performed with the post-hoc Bonferroni test to compare and replicate averages per line, as well as a one-way ANOVA variance analysis with Tukey test for multiple comparisons, to compare all columns, with a limit of significance set to 0.05 (*GraphPad Prism 7.02 Software*).

## 6.4 Results

There were no significant postoperative complications among the studied individuals, with the exception of two rabbits, to which the local edema persisted for three weeks. The rabbits have completely regained the function of the affected limb at 4 weeks.

### 6.4.1 Macroscopic evaluation

At the time of sacrifice, no signs of local inflammation, infection or other joint pathological aspects were observed. The overall appearance seemed better for group B, with a clearer coverage of the defect with a more likely cartilage tissue (Figure 6.5).

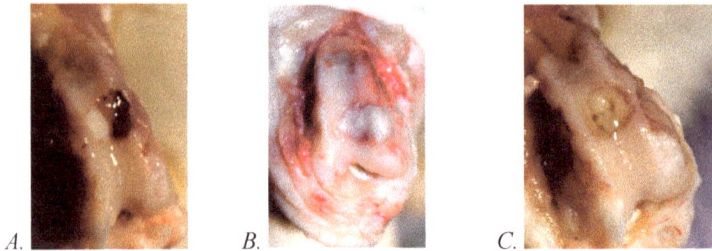

*Fig. 6.5. Macroscopic aspect.*

***A.** Group A – control. **B.** Group B – LAF. **C.** Group C – PLA.*

The overall macroscopic evaluation of cartilage repair is presented in Table 6.3 and Figure 6.6. The highest value of the Wayne score was obtained in the LAF group, with the average of 11.3 out of 16, and the ANOVA test showed a statistically significant difference between groups ($p = 0.0041$).

***Table 6.3.** Macroscopic evaluation of cartilage repair according to Wayne Score*

| | Group A – No ADSC | | | Group B – LAF | | | Group C - PLA | | |
|---|---|---|---|---|---|---|---|---|---|
| **Rabbit no.** | 1 | 2 | 3 | 1 | 2 | 3 | 1 | 2 | 3 |
| **Individual score** | 6 | 8 | 7 | 12 | 10 | 12 | 10 | 12 | 11 |
| **Group score** | 7 | | | 11,3 | | | 11 | | |

The Tukey Multiple Comparison Test showed statistically significant differences between Group A and B, and respectively C, but without significant differences between Groups B and C (Table 6.4).

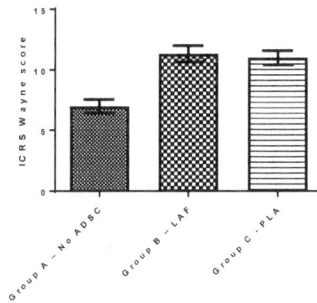

**Fig. 6.6.** *Total Wayne Score for the three groups*

**Table 6.4.** *The results of statistical analysis between the three groups*

| Tukey Multiple Comparison Test | Mean difference | q | Significant? p < 0.05? | Summary | 95% CI |
|---|---|---|---|---|---|
| **Group A vs Group B** | -4.333 | 7.120 | **Yes** | ** | -6.974 to -1.693 |
| **Group A vs Group C** | -4.000 | 6.573 | **Yes** | ** | -6.641 to -1.359 |
| **Group B vs Group C** | 0.3333 | 0.5477 | No | ns | -2.307 to 2.974 |

** *statistically significant p < 0.05*

Macroscopic evaluation showed significantly higher cartilage surface coverage (p <0.001) and significantly smoother area (p <0.05) in group B vs. group A, cartilage coverage statistically significantly better (p <0.001) and also the color (p <0.05) in group C vs. group A (Figure 6.7).

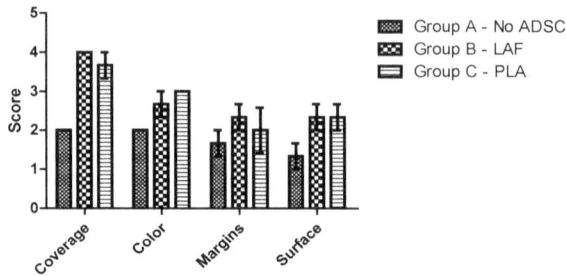

*Fig. 6.7. Wayne score parameters comparison between the three groups.*

### 6.4.2   Histological evaluation

As it can be seen in Figure 6.8, the cartilaginous tissue appears bright blue in Alcian Blue staining. The difference between cartilaginous and bone tissue is remarkable. Cartilaginous tissue is rich in chondrocytes and matrix. The rounded form of chondrocytes, gathered in groups of two or more, is observed in a granular or almost homogeneous matrix.

We performed a qualitative analysis of the repaired tissue in the two stem cell derived groups derived from adipose tissue using specific histological parameters (Figure 6.9). According to Group B assessment, we obtained the following:

- color, thickness, alignment of the neocartilage were similar in the two groups, with almost complete bonding of the native cartilage and continuity tidemark

- hyaline cartilage more abundant, intact and more regular than in the PLA Group

- lower cellularity, less disorganized, with fewer chondrocytes agglomerates than the PLA group

- better reconstruction of the subcondral bone with less fibrous tissue and almost complete adhesion to the adjacent bone.

These images confirm gene expression analysis and macroscopic evaluation, showing the remarkable potential of LAF cells to differentiate into cartilage tissue.

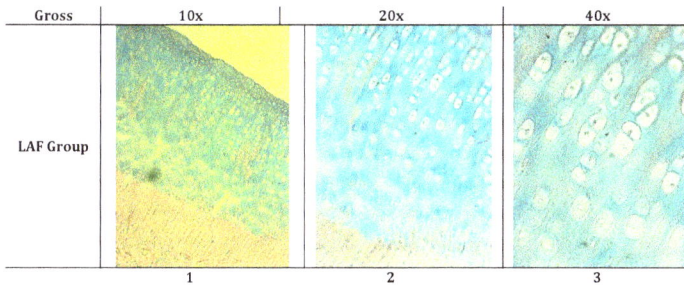

***Fig. 6.8.*** *1-3. Microscopic aspects of regenerated articular cartilage in LAF Group (B).*

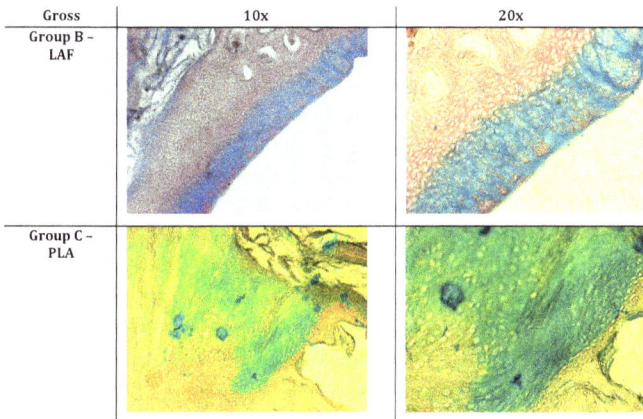

***Fig. 6.9.*** *Comparative microscopic aspects of articular cartilage formation in the two groups with ADSCs (Acian Blue staining).*

### 6.4.3 Analysis of gene expression

According to the methods described above (real time PCR quantitative - qPCR), we noticed that expression of genes involved in cartilage differentiation was significantly stronger in group B with LAF. More explicitly, in group B the significance level of ACAN gene expression was the highest, compared to groups A and C ($p < 0.00051$). Regarding the expression of COL2A1 and SOX-9 genes, it was statistically highly significant compared to group A (p<0.0005) and statistically significant compared to group C (p <0.05) (Figure 6.10).

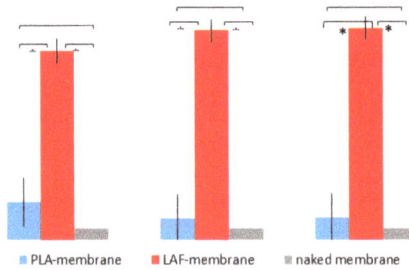

***Fig. 6.10.*** *Comparative expression of genes involved in cartilage differentiation*
*(\* p<0.05; \*\* p<0.005; \*\*\* p<0.0005).*

### 6.4.4   MicroCT scan

We performed an analytical comparative analysis of CT images obtained from groups B and C (Figure 6.11). We have observed almost total restoration of the thickness of the articular cartilage layer and subcutaneous bone trabecularity, slightly more pronounced in the LAF group, in agreement with the microscopic sections in Figures 6.8 and 6.9.

### 6.5   Discussions

First, the results of our study show that, in an animal model, surgical treatments using collagen matrixes impregnated with mesenchymal stem cells can repair osteochondral defects in the knee joint. But the most important discovery is that, in terms of regeneration of articular cartilage, stem cells derived from adipose tissue by minimal processing of the liposacral fluid fraction (LAF cells) can be at least as useful as stem cells obtained from laboratory processed lipoaspirate (PLA cells). This statement is supported by the following results: the best Wayne macroscopic score obtained by the LAF group (11.3 points), the significantly stronger expression of cartilage proliferation genes in the LAF group (qPCR), and very good microCT and histological evaluation of regenerated cartilage and subchondral bone.

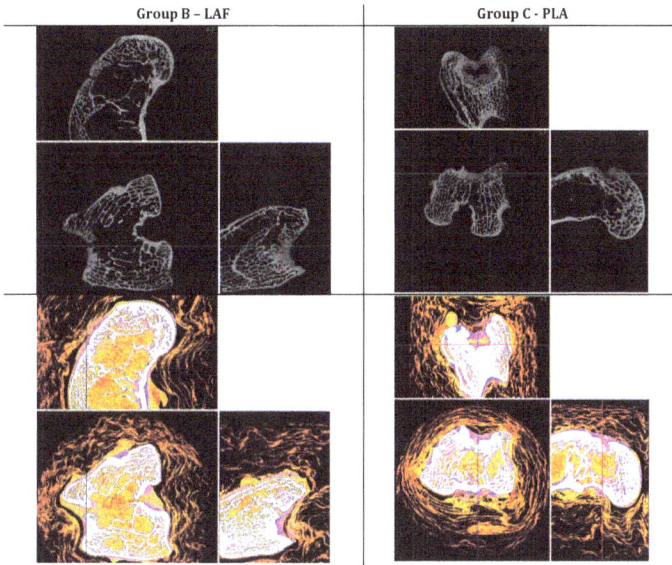

***Fig. 6.11.*** *Comparative microCT scan aspects for LAF and PLA Groups*
*(lesions are marked at the crossing of coloured lines).*

Our study was built as a pilot research and we are aware of some limitations. Although the rabbit is often used as a small animal model, the transposition into the human model may not be as clear. Also, the small number of animals included did not allow us to draw statistically significant conclusions.

Stem cells derived from adipose tissue (ASC) can be isolated in similar quantity and quality, both from the liquid and the fatty fraction of the lipoaspirate[210,216]. The purpose of this research was not to describe the process of isolation and cultivation of these cells, because the whole process the whole process has already been exposed[208,209], but to provide practical evidence of their potential for *in vivo* use for osteochondral regeneration.

Osteochondral regeneration from stem cells derived from adipose tissue implanted on various matrices has been described in the literature in various *in vivo* experiments[215,217]. But in almost all cases, ASCs are obtained by processing the stromal-vascular fraction of lipoaspirate. Reports on the use of LAF for *in vivo* experiments are rare[218], and those about cartilage repair are absent. In this context, our study is a novelty, and the results encourage us to continue our research.

The role of cells in the regeneration of cartilage is undeniable. As our results show, the groups with added ASCs presents statistically significantly better cartilage formation than the control group, where the type I/III collagen membrane has been implanted over the subchondral microfractures, although there are very good results with this method in literature[219]. The use of a validated condrogenic matrix (*Chondro-Gide*[R], *Geistlich Pharma AG*)[220] eliminated the risk of bias due to surgical procedure and implant.

It is remarkable the significantly higher expression of chondrogenic genes in the LAF group, which may mean that their activation, differentiation and proliferation capabilities are much stronger than those of PLA cells. This is probably due to the large number of trophic molecules and tissue fractions contained in LAF cell fluid that co-operates with stem cells to exploit regenerative properties[209] and, as other studies suggest[220], the process of isolating PLA cells by collagenase digestion would reduce their performance. Although the overall result of the LAF cell group appears to be slightly better compared to the PLA group, the most important advantage is the relative ease of harvesting with minimal manipulation[203,209,218,220].

## 6.6    Conclusions

• ASCs are valuable options for osteocondral repair.

• The stem cells derived from LAF appear to have a better activity and effects compared to the PLA cells, as regards the repair of the cartilage, being aided by the trophic molecules in the liquid.

• LAF cells can be quickly separated with minimal tissue manipulation, so they are cheaper and suitable for single-stage surgical procedures.

## 7.     Study 3. Comparative assessment of the healing of focal lesions of articular cartilage on an animal model, by using stem cells from the iliac crest versus stem cells from adipose tissue

### 7.1     Introduction

Currently, the pathology and treatment of focal cartilage lesions are an intense field of research, both at the level of laboratory as well as *in vivo*. The limited intrinsic healing capacity of cartilaginous tissue is due to the small number of specialized cells with low mitotic activity and the absence of blood vessels and undifferentiated potent cells.

There are currently several therapeutic procedures seeking to repair cartilage lesions, none of which have the maximum desired efficiency, the ultimate goal being the healing of lesion by forming a new hyaline cartilaginous tissue with structure and functionality identical with the healthy one. Due to the difficulty, the high cost, legal and ethical issues of chondrocyte cultivation and implantation methods, it is attempted to develop methods based on mesenchymal stem cells *(MSC)* as a key element of cartilage regeneration.

### 7.2     Work hypothesis

The purpose of the study is to evaluate the efficiency of methods using multipotent mesenchymal stem cells (MSC) to heal focal joint cartilage defects.

The overall objective is to perform a comparative experimental preclinical study in vivo on ovine animal model for the repair of focal cartilage defects by methods involving the use of MSC.

The first objective consists in carrying out methods of harvesting stem cells from concentrated aspirate from the iliac crest (BMC – Bone Marrow Concentrate) and from adipose tissue (ASC – Adipose derived Stem Cells) on sheep model and characterizing the obtained cell population.

The second objective is to develop interventional methods on 15 ovine specimens by creating cartilage defects to the subchondral bone at the central portion of the medial femoral condyle of the left knee and repairing them by specific techniques involving a new type of collagen scaffold and MSC derived from BMC and ASC, or respectively without added cells.

The third objective is to perform a comparative clinical, anatomo-pathological and imagistic analysis of the evolution of focal lesions of articular cartilage after 7 months from the repair interventions with the involvement of stem cells.

As an element of originality, the study will compare in the same experiment the three methods that rely on MSC. The study will help deepen the understanding of MSC's role in the treatment of articular cartilage lesions, will test *in vivo* the new type of scaffold, and will allow to improve the possibilities of interventional therapy by developing methods with minimal cost and morbidity and maximum efficiency. It will also serve as a basis for diagnostic and treatment algorithms of cartilage lesions in the current practice.

## 7.3  Materials and methods

### 7.3.1  Ethical considerations

The study was conducted with the approval of the Ethics Committee of the University of Medicine and Pharmacy "Iuliu Haţieganu" Cluj-Napoca, approval no. 237/19.06.2014, and experiments were conducted in compliance with EU Directive no. 63/2010 and law no. 43/2014 in Romania.

### 7.3.2  Animals and experimental environment

For this study, 15 adult female sheep from the Turcana breed were used, coming from the same life environment, aged between 24 and 36 months (average 30 months ± 6 months), average weight 45 kg (±5 kg), without apparent health problems detected by the clinical examination performed by a veterinarian.

Surgical interventions were carried out in the operating theater of the Surgery Clinic within the USAMV - University of Agricultural Sciences and Veterinary Medicine Cluj-Napoca by a mixed team of veterinary surgeons, orthopedic surgeons from the University of Medicine and Pharmacy "Iuliu Haţieganu" Cluj-Napoca, assisted by biologists specialized in working techniques with autologous biological products, especially mesenchymal stem cells. The surgical interventions were performed under general anesthesia and under strict aseptic conditions specific to the operating theaters.

The preoperative preparation consisted in the separation of sheep from the herd and their acclimatization in quarantine for 14 days preoperatively. The day prior to surgery, the animals were subjected to a dietary alimentation for solids, but without restrictions for water. Prior to induction of anesthesia, local preparation was carried out by trimming the incision region.

Postoperatively, the animals were housed in a shelter of 50 sqm. per straw layer, isolated from other animals and protected from weathering, providing natural and artificial light in addition, access to drinking water, fodder and concentrated forage, as well as additional intake of NaCl salt (Figure 7.1).

### 7.3.3    Group randomization

The 15 animals were allocated in three groups using a random number generating software. The list of assigned codes was retained by a member of the team and was not disclosed to the investigators until the samples had been evaluated.

*Fig. 7.1. Housing conditions.*

**Group A** (nA = 6, sheep numbered 1 to 6) included sheep which received stem cell treatment from fatty tissue concentrate (ASC), namely the stromal-vascular fraction (SVF - Stromal Vascular Fraction), unprocessed (without enzymatic or other multipotent stem cell separation) mixed with platelet rich plasma (PRP - Platelet Rich Plasma).

In **Group B** (nB = 6, sheep numbered 7 to 12), the cells were obtained by concentrating the bone marrow aspirate from the iliac crest, activated with Batroxobin (*Plateltex Act*[R], *Plateltex SRO*), not being mixed with PRP.

Control **Group C** (nC = 3, sheep numbered 13 to 15) did not receive stem cell treatment, only the local cells present in the created defect, without adding concentrated aspirate taken separately from the adipose tissue or haematogenous bone marrow.

In all cases, the same type of collagen I/III implant was used as support for the studied cell populations and its fixation in situ was performed with fibrin adhesive obtained either from a commercial preparation (*Tisseel Lyo*[R], *Baxter*), or by activating PRP with Batroxobin (*Plateltex Act*[R], *Plateltex SRO*), alternatively.

### 7.3.4    Preoperative preparation

For each animal a peripheral venous catheter approach was performed on the antebrachial cephalic vein and an intravenous infusion of 0.9% saline was installed for each. After the anesthesia, 20 ml of blood was taken to obtain PRP and to carry out biological samples of interest (complete blood count with platelet counts from both natural plasma and PRP).

Anticholinergic premedication consisted of subcutaneous administration of 1% Atropine at doses of 0.01 mg/ kg body weight to prevent the hypersalivation or severity of severe bradycardia produced by anesthetics (α-2 adrenergic agonists).

After 10-15 minutes from the administration of Atropine, sedation or neuroplegia was achieved by intravenous administration of Xylazine (*Narcoxyl 2%*) at a dose of 0.02 to 0.05 mg/kg body weight (an α-2 adrenergic agonists with sedative or analgesic effects depending on the dose). For anesthesia induction via endotracheal intubation was administered i.v. Ketamine (*Ketamidor 10%*) at 2-5 mg/kg body weight dose. For endotracheal intubation, 25 mm diameter catheters were used and the narcotic induction was made with Isoflurane (*Forane*) at a concentration of 3-4%, and the maintenance of narcosis with a concentration of 1-2% of Isoflurane.

Intraoperative analgesia was provided by the administration of Ketamine microdoses (1 mg/ml) in continuous infusion or bolus (Figure 7.2.A). Intraoperative monitoring included pulse, respiration, temperature, ECG and $SpO_2$ values.

Further, the animals were positioned on a standard operative table in ipsilateral decubitus with the exposure of the areas of interest (as the left knee in each subject, as well as the harvesting areas for stem cells – the postero-superior iliac crest or the dorsal region of the tail, as appropriate), followed by regional aseptization by alcoholic betadine solution and isolation of the surgical field (Figure 7.2.B), according to standard protocols, using disposable single use self-adhesive waterproof surgical drape sets (*Opero*[R] *SET, Mercator Medical*).

Control of postoperative analgesia was achieved by the single administration of Flunixin Meglumine (*Fynadine*) 2.2 mg/kg body weight. Intramuscular antibiotic prophylaxis was performed for 3 days postoperatively by administering Cefalexin (*Solvasol*) at a dose of 10 mg/kg body weight, with continuous monitoring of physiological parameters.

**Fig. 7.2.A.** *General anesthesia.*

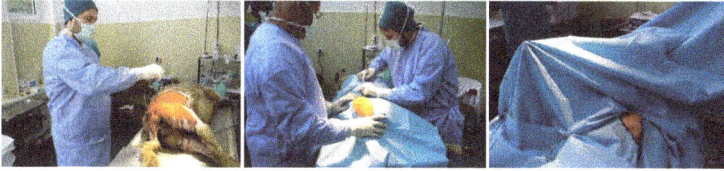

***Fig. 7.2.B.*** *Sterile draping.*

### 7.3.5    Harvest, separation, concentration and characterization of stem cells

### 7.3.5.1 BMC harvesting

With the anaesthetized sheep positioned in the left lateral decubitus and isolated surgical field, the right postero-superior iliac crest was punctured by a special bone trocar (3 mm $\varnothing$ x 11 mm L) from the Concemo® kit (*Proteal Bioregenerative Solutions S.L., Spain*), for BMC sampling (Figure 7.3). Care was taken to penetrate between the bones boundaries, from where it was softly aspirated, without forcing the vacuum, and with the permanent rotation of the trocar, in order to avoid clogging the suction holes (Figure 7.4).

***Fig. 7.3.*** *Concemo® Kit for obtaining BMC*
*(http://www.proteal.com/en/ortho-products-proteal/concemo-proteal).*

All materials used for bone marrow harvesting and collection were washed with heparin solution at a concentration of 2000 IU/ml to prevent premature clotting of the aspirate. Another 10 ml of medullary aspirate was collected which was filtered (to remove fat and residual bone tissue) and placed in a Concemo® collection tube in which 1 ml of sodium citrate 3.8% (existing in the kit) was injected in advance. The liquid was gently mixed and introduced into the Duromter II® Cellular Concentration Unit for centrifugal processing to obtain 1.2 ml of BMC.

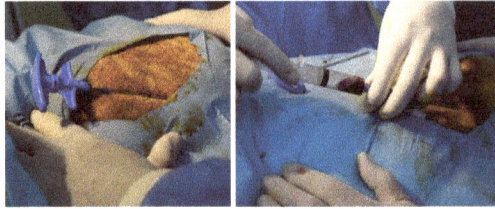

***Fig. 7.4.*** *Medullar aspirate extraction from postero-superior iliac crest.*

Following several determinations, the centrifugation protocol was validated at 2500 rpm for 15 minutes (based on the study by Gobbi A. et al in 2011), whereby the concentration of CD44+ multipotent mesenchymal cells was successfully doubled (Table 7.1).

From the amount of BMC obtained (approximately 1.2 ml), 0.5 ml was mixed with 5:1 ml of Plateltex Act® solution, thereby activating and gelling the concentrate solution, thereby favoring adhesion to the collagen scaffold (Figure 7.5).

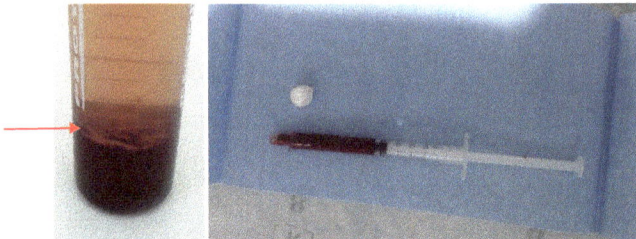

***Fig. 7.5.*** *Bone marrow concentrate (obtained from the separation layer between plasma and figured elements) and the collagen implant to be impregnated.*

The remaining amount of BMC was sent to the USAMV Cell Biology Laboratory where marking with anti-CD44 antibody was performed, reagent for the isolation and characterization of ovine mesenchymal stem cells (Anti-CD44 antibody [Hermes-1] ab119335 and mouse monoclonal 2A 8F4 Anti-Rat IgG2a heavy chain (FITC) ab99665, *Abcam UK*).

Flow cytometry techniques determined the proportion of CD44+ cells in BMC mononuclear cell population (basically the mesenchymal stem cell population). At the same time, a native, unprocessed bone marrow aspirate was sent out of which the same percentage of CD44+ cells was determined, the ratio of the two values being the concentration factor of mesenchymal stem cells in BMC (Table 7.1).

***Table 7.1.*** *BMC CD 44+ cells distribution in Group B*

| Sheep no./ Group B | CD44 concentration in unprocessed aspirate | CD44 concentration in BMC (processed) | Relative concentration increment | Activated BMC volume injected in the scaffold (ml) | Plateltex lot no. |
|---|---|---|---|---|---|
| Sheep 7 | 13,8% | 37,5% | 2,72x | 0,5 | PT18041401 |
| Sheep 8 | 12,0% | 27,6% | 2,3x | 0,5 | PT18041401 |
| Sheep 9 | 25,4% | 50,4% | 1,98x | 0,5 | PT18041401 |
| Sheep 10 | 14,6% | 30,3% | 2,07x | 0,5 | PT18041401 |
| Sheep 11 | 9,4% | 18,5% | 1,97x | 0,5 | PT18041401 |
| Sheep 12 | 19,0% | 39,4% | 2,07x | 0,5 | PT18041401 |

## 7.3.5.2 ASC harvesting

For stem cell isolation (ASC), adipose tissue from the tail region was harvested using St'rim® sampling kits (*THIEBAUD S.A.S., Paris France*). In advance, the region was infiltrated with the tumescent solution, prepared in situ from 0.9% saline, 1% lidocaine, 1% epinephrine, and sodium bicarbonate. This solution has a dual role, vasoconstrictor to prevent bleeding and fatty tissue fluidization in order to be better sampled.

After local disinfection with betadine and waiting approximately 20 minutes after injection, 20 ml of fresh adipose tissue was taken using the special cannula kit (Figure 7.6). After centrifugation, about 1 ml of stromal-vascular fraction (SVF) concentrate (Figure 7.7), which is rich in inactive multipotent cells, and mixing with PRP (Platelet Rich Plasma) (Figure 7.8) is required, which contains activating factors.

Part of the fatty tissue concentrate was sent to the molecular and cellular biology laboratory for the flow cytometric determination of CD44 + mononuclear cells.

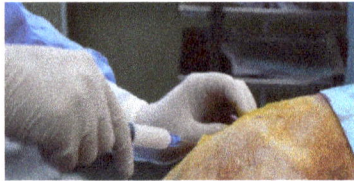

***Fig. 7.6.*** *Adipose tissue withdraw.*

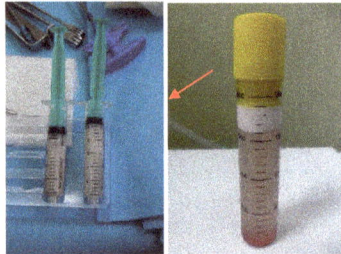

***Fig. 7.7.*** *Adipose tissue before and after concentration.*

### 7.3.5.3 PRP preparation

From 20 ml of blood sample taken after anesthesia has been induced, 10 ml were processed with the special OrthoPras® kit (*Proteal Bioregenerative Solutions S.L., Spain*), after centrifugation, 1 ml of PRP concentrate was then mixed with the adipose tissue concentrate.

***Fig. 7.8.*** *PRP Concentrate.*

### 7.3.6   Preparation of implants

We proposed to test a new type of osteochondral implant of porcine collagen I/III, derived from a biphasic bone implant substitute containing collagen I/III and hydroxyapatite, commercially available demineralized MatriBone Ortho® (*Biom'Up,*

*France*). The new implant was obtained by decalcification of the biphasic implant. Using a biopsy punch of 8 mm diameter, collagen I/III implants were processed under sterile conditions to perfectly fit the created osteochondral defect.

The amount of liquids thus obtained (approximately 1 ml) was injected into the collagen implant until saturation (Figure 7.9), and the rest at the osteocondral defect site. The implant was inserted to the defect immediately afterwards.

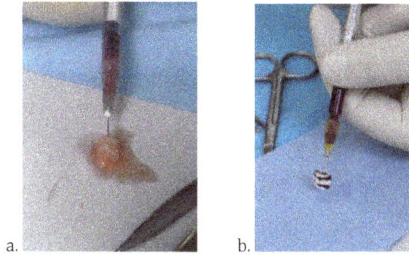

**Fig. 7.9.** *Implant injection: a) SVF + PRP mixture; b) BMC aspirate.*

### 7.3.7 Surgical technique

After the preparation of the surgical field with sterile disposable drapes (*Opero*[R] *SET, Mercator Medical*), a medial parapatellar approach was performed on the anterior left knee (Fig. 7.10 a). After the incision of the skin and the subcutaneous connective tissue, lateral dislocation of the patella was performed evidencing the medial femoral condyle.

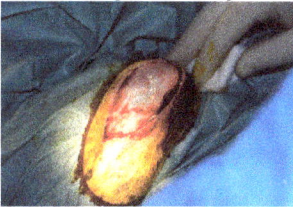

**Fig. 7.10 a.** *Medial parapatellar approach of the knee.*

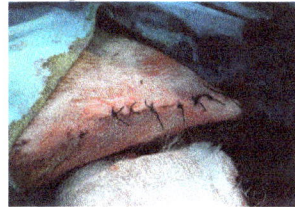

**Fig. 7.10 b.** *Wound closure.*

The osseocartilaginous defect was induced by an 8 mm diameter drill bit with a depth of 4 mm until a local hemorrhage was observed in the subchondral bone (Outerbridge grade IV). After the defects were performed, their reconstruction was performed, depending on the group, then wound suture was made in planes and local sterile compress bandages were applied (Figure 7.10 b).

Different treatment methods were used: thus, in the first study group, the collagen matrix I/III impregnated with a combination of stem cells concentrate derived from adipose tissue ASC and PRP thrombocyte concentrate activator was applied to the cartilage defect area, all covered with a special fibrin glue (*Tisseel Lyo*[R], *Baxter, Deerfield, IL, USA*) (Figure 7.11 d).

In the second study group, the collagen scaffold was impregnated with the multipotent cell concentrate from the iliac crest, and the fixation was done with the same type of biological adhesive. In the control group, in the created cartilaginous defect only the collagen matrix was inserted and fixed without addition of cells.

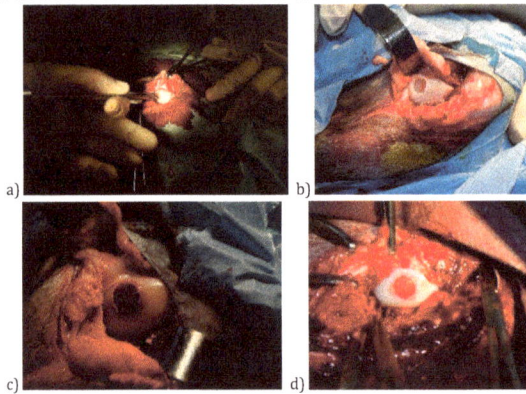

***Fig. 7.11****. a). Defect creation; b). Osteochondral defect: c) The implant in place; d) Implant sealed with fibrin glue.*

The procedures of creating and repairing of condral defects, sampling and preparation of cell aspirates, were performed in a single intervention, under conditions of maximum sterility in accordance with surgical standards.

The postoperative management protocol was identical for all animals. Postoperatively, they were housed and cared for under the best conditions, avoiding exposure to risks in the USAMV Cluj-Napoca Biobase. The animals were monitored weekly by investigators for bandaging, clinical evolution follow-up, and the occurrence of any adverse effects or complications.

### 7.3.8   Criteria for evaluating results

At seven months after the experiment, the animals were euthanized by administering an intravenous potassium chloride injection after a prior anesthesia (the anesthetic protocol described above). The distal femur was dissected and taken to be further subjected to a radiological examination (4 cases), MRI (4 cases) or CT (4 cases). After these imaging investigations were performed, samples were taken with a mini-oscillating saw from the lesion area and immediately fixed in a 10% formaldehyde solution for histological and immunohistochemical studies.

The objectives of the study were: macroscopic and microscopic histological evaluation of the quality of the repair tissue, observation of internal morphology and microstructure through high performance imaging investigations (radiography, CT and MRI) and an immunohistochemical study for the presence of collagen type I and II.

### 7.3.8.1 Macroscopic evaluation

Immediately after sacrifice and dissection, a first macroscopic assessment was performed according to the Wayne Cartilage Repair Score[227] (Table 6.1 - Chapter 6). This scoring system comprises a 16-point scale based on four parameters: coverage of the defect with regenerative tissue, the color of the neocartilage, edges, smoothness and the color of the defect.

### 7.3.8.2 Microscopic evaluation

The histological and immunohistochemical samples were processed and interpreted at the Department of Pathological Anatomy of UMF Cluj-Napoca.

After a 24 hour incubation with 10% formaldehyde buffer in PBS at 4°C and pH 7.5, the samples were washed in water for one hour and then decalcified for 15 days (*Decalcifier II, Leica*), according to the manufacturer's instructions, with daily change, dehydrated using ethyl alcohol and methyl alcohol, clarified with xylene, sunken in paraffin and molded.

The tissue samples were then cut into 15 μm thick sections using a Leica SM2000R microtome (*Leica Biosystems, Wetzlar, Germany*). Subsequently, several stains were used to highlight cartilage and bone tissue: hematoxylin eosin (HE), Masson's trichrome, Red Sirius, Safranin O and Fast Green.

In the HE coloration, we can see the following aspects: dark blue nuclei, pink-red cytoplasm, vivid red blood cells, pale pink collagen and pink elastic fibers. Masson's trichrome coloring provides contrast details and highlights collagen. We will also notice: gray-red nuclei, pink-violet cytoplasm, blue connective tissue, intense blue bone, and intense red osteoid. Red Sirius staining will show us the blue nuclei and red connective tissue. Alcian Blue staining has tropism for glycosaminoglycans in cartilaginous tissue, which it highlighted in blue-green. The coloration of Safranin O allows us to visualize the proteoglycans of the articular cartilage in red and the nuclei in blue on a green background.

The quantitative microscopic evaluation was based on the modified O'Driscoll score[228] (Table 7.2).

### 7.3.8.3 Immunohistochemical evaluation

We conducted an immunohistochemical qualitative study in order to detect the presence of collagen type II from sheep (which are normally present in the hyaline articular cartilage) and type I collagen fibers (present at the level of the porcine collagen implanted and in the fibro-cartilaginous repair tissue). The hypothesis of hyaline tissue repair involves the predominance of collagen type II ovine fibers and a low proportion of collagen type I.

***Table 7.2.*** *Modified O'Driscoll score (after Veronesi et al[221])*

| Tissue | Parameters | Points | | | | |
|---|---|---|---|---|---|---|
| | | **0** | **1** | **2** | **3** | **4** |
| Cartilage Min-max 0-31 | Tissue morphology | Fibrous tissue | Mostly fibrocartilage | Mix Hyaline and Fibrocartilage | Mostly hyaline cartilage | Hyaline cartilage |
| | Matrix staining | None | Slight | Moderate | Normal | |
| | Surface regularity | Disrupted surface | Fissuring | Horizontal lamination of surface | Smooth and intact | |
| | Structure integrity | Severe disintegration | Slight break | Normal | | |
| | Neo-cartilage alignment with native cartilage | Depressed | Elevated | Leveled | | |
| | Thickness of neo-cartilage | No cartilage | 50%-100% of normal cartilage | Similar to normal | | |
| | Bonding to native cartilage | Not bonded | Partially bonded at both ends or bonded at one end | Bonded at the both ends | | |

| | | | | | | |
|---|---|---|---|---|---|---|
| | Chondrocyte clustering | 25%-100% of the cells | <25% of cells is grouped in clusters | No clusters | | |
| | Hypocellularity | Severe | Moderate | Slight | Normal | |
| | Cell distribution | Disorganized | Moderate disorganization | Slight disorganization | Columnar | |
| | Degenerative changes in the adjacent native cartilage | Severe hypocellularity, poor or no staining | Mild or moderate hypocellularity, slight staining | Normal cellularity, mild clusters, moderate staining | Normal cellularity, no clusters, normal staining | |
| | Tidemark continuity | Absent | Partial | Present | | |
| **TOTAL** | | | | | | |
| **Bone Min-max 0-12** | Reconstruction of subchondral bone | No reconstruction | Minimal | Reduced | Normal | |
| | Bone infiltration into area defect | None | Partial | Complete | | |
| | Bonding with adjacent bone | Without continuity on either edge | Partial on both edges | Complete on one edge | Complete on both edges | |
| | Subchondral bone morphology | Only fibrous tissue | Compact bone and fibrous tissue | Compact bone | Trabecular with some compact bone | Normal trabecular bone |
| **TOTAL** | | | | | | |
| | Report mineralized bone/osteoid | | | | | |

The immunohistochemical staining technique was performed by the indirect three-stage method of highlighting with DAB of collagen type I and II using sheep specific anti-collagen antibodies. We used the primary anti-collagen I antibody Collagen I alpha I

Antibody NBP1-30054 (*Novus Biologicals, LLC, USA*) and the primary anti-collagen II antibody Collagen Antibody PA1-26206 (*ThermoFisher Scientific, USA*).

The histological blades obtained were examined and photographed with a Leica microscope equipped with an ICC50HD acquisition chamber.

### 7.3.8.4 Imagery

*The radiological examination* was performed with a Roentgen CS3000 equipment, at USAMV Cluj-Napoca, through a latero-lateral and cranio-caudal exposure of the limb.

*The Computer-Tomography (CT)* scan was performed with a Siemens Somatic Scope device at USAMV Cluj-Napoca. Coronal and transverse cross sections of the limb were made.

*The Magnetic Resonance Imaging (MRI)* evaluation was performed with a Bruker Biospec 7.0 Tesla apparatus equipped by the National Magnetic Resonance Center (CNRM), within the Faculty of Physics of Babeş-Bolyai University[222].

The Bruker Biospect 70/16 USR instrument, having a magnetic field intensity of 7.04 Tesla, has the ability to provide, non-invasive and non-destructive, a combination of functional and anatomical information in vivo. State-of-the-art MRI CryoProbe™ technology, combined with USR magnets, offers a high spatial resolution (by the order of tens of microns). Allows imaging and spectroscopy on 1H and a series of other MR-active nuclei, such as 31P, 23Na, 19F and 13C.

Acquisition Protocol: The MRI investigation was conducted to obtain *2D* anatomical images with the *2D* TURBO-RARE Acquisition Protocol[223] (TURBO-RARE Acquisition Protocol), which allowed morphological information to be obtained from the area of interest. Various acquisition parameters were used that resulted in RM imaging, both weighted T1 (spin-lattice relaxation time) and weighted T2 images (spin-spin relaxation time).

The TurboRARE protocol is a Spin Echo protocol used extensively in both preclinical and clinical trials; this acquisition protocol provides good quality images (high resolution) in a short time. Anatomical information were obtained by weighted T2 (T2-W) images in all three orthogonal planes: axial, coronal, and sagittal.

Another MRI acquisition protocol (Magnetic Resonance Imaging) used in this study was FLASH (Fast Low Angle Shot[224]). This is an echo-gradient protocol that allowed observation of the area of interest with another contrast variation, necessary to accurately identify the boundaries of the bone implant.

Only the acquisition of two-dimensional images was based on several factors; the most defining factor that prompted the acquisition of *2D* images was the size of the volume to be scanned, which had dimensions of the order of $cm^3$, which for *3D* acquisition the major disadvantage is the scanning time, which can reach an acquisition interval of order of hours to obtain an image resolution in the range of tens of micron. Image acquisition parameters are shown in Table 7.3.

***Table 7.3.*** *MRI images acquisition parameters*

| No. | Acquisition protocol | TR (ms) | TE (ms) | FA ($^O$) | TA (min) |
|-----|----------------------|---------|---------|-----------|----------|
| 1 | RARE-T1 | 2217.7 | 12.3 | 180 | 17 |
| 2 | FLASH | 587 | 5.4 | 40 | 11 |
| 3 | Turbo-RARE-T2 | 3838.9 | 33 | 180 | 2 |
| 4 | TurboRARE-high-res | 2500 | 36 | 180 | 10 |
| 5 | TurboRARE-*3D* | 1500 | 45 | 180 | 13 |

### 7.3.8.5 Data processing and analysis

The statistical interpretation was based on the Student Test for independent double-tail variables. The significance limit was established as $p < 0.05$.

### 7.4    Results

The postoperative evolution of animals went normally, with immediate mobilization, within the limits of pain. There were no major complications, except 2 collected seromas at the incision level, in 2 animals, which were evacuated by puncture and transcutaneous drainage, with favorable evolution (Figure 7.12). At 3 weeks the wounds were healed. The limping decreased considerably between 4-6 weeks postoperatively.

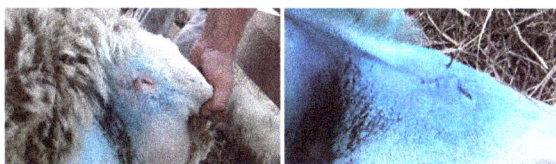

***Fig. 7.12.*** *Healing of the seroma, without drainage*

At 7 months postoperatively, prior to sacrifice, the animals were clinically healthy and showed no signs of limping; the surgical wound has been completely healed, with normal scarring.

### 7.4.1 Macroscopic and microscopic evaluation

The results recorded in the Wayne Macroscopic Score Analysis are shown in Figure 7.13. There were no statistically significant differences between the three groups of animals. The averages and standard deviations of the groups are shown in Table 7.4.

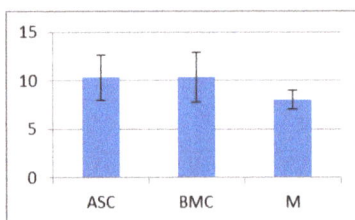

**Fig. 7.13.** *Scores comparison between groups.*

**Table 7.4.** *Means and standard deviations of macroscopic score between the three groups.*

| Macro Wayne | ASC | BMC | CONTROL |
|---|---|---|---|
| STD. DEVIATION | 2,338090389 | 2,581989 | 1 |
| MEAN | 10,33333333 | 10,33333 | 8 |

The results of the microscopic histological evaluation are presented in Figure 7.14, and the mean and standard deviations of the groups in Table 7.5.

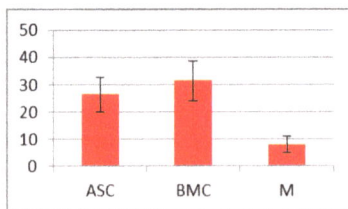

**Fig. 7.14.** *Microscope score comparison between groups.*

***Table 7.5.*** *Means and standard deviations of microscopic score between the three groups.*

| Micro total | ASC | BMC | CONTROL |
|---|---|---|---|
| STD. DEVIATION | 6,186005712 | 7,420691792 | 3 |
| MEAN | 26,33333333 | 31,33333 | 8 |

The microscopic score was higher for the BMC group, with an average of 5.2, compared to the ASC group (mean 4.38) and the control group (mean 2.66). Differences have been shown to be statistically significant between all three groups both as a general score and in part for individual cartilage and bone evaluation scores (Table 7.6), with the exception of cartilage assessment between groups A and B.

***Table 7.6.*** *"p" values at comparative analysis of microscopic scores.*

| Student Test | Micro total | Micro cartilage | Micro bone |
|---|---|---|---|
| ASC vs BMC | 0,015* | 0,064 | 0,004* |
| ASC vs M | 0,001* | 0,000* | 0,047* |
| BMC vs M | 0,000* | 0,000* | 0,012* |

*p<0.05 statistically significant

The results shown in Figure 7.15 were obtained when assessing the microscopic average scores. Related to cartilage regeneration, and mean bone regeneration results are shown in Figure 7.16.

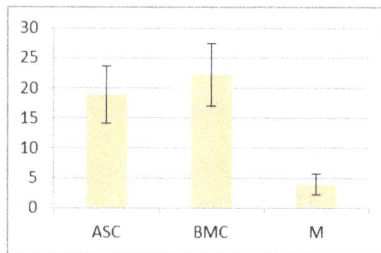

***Fig. 7.15.*** *Comparison of microscopic cartilage regeneration score between groups.*

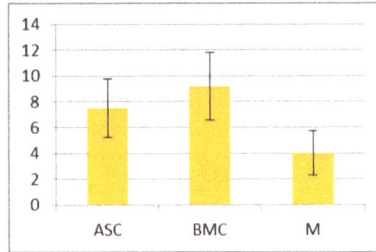

***Fig. 7.16.*** *Comparison of microscopic bone regeneration score between groups.*

Following the interpretation of microscopic scores for evaluation of cartilaginous and bone areas, some statistically significant differences were obtained between ASC and BMC groups, which are shown in Table 7.7.

***Table 7.7.*** *Comparison of microscopy score parameters values at the cartilage (first 4 layers) and bone levels (5$^{th}$ layer), between ASC and BMC groups, with statistical significance.*

|  | ASC Group mean | BMC Group mean | "p" value |
|---|---|---|---|
| **Morphology** | 2,00 | 3,00* | 0,012* |
| **Matrix staining** | 1,67 | 2,83* | 0,034* |
| **Cell distribution** | 1,33 | 2,17* | 0,042* |
| **Tidemark continuity** | 1,00* | 0,17 | 0,004 |
| **Total microscopic score - bone** | 7,50 | 9,17* | 0,004* |

*p<0.05 statistically significant

Statistically significant differences were recorded between the stem cell treated groups (A and B) and the non-stem cell control group, at all microscopic score parameters, except for the thickness of the neocartilage, the degenerative changes in the adjacent native cartilage, the reconstruction of the subchondral bone and connection with the adjacent bone (Table 7.8). The mean values for the treatment groups were higher than in the control group.

**Table 7.8.** *Comparison of microscopic scores at the level of cartilage and bone for Groups ASC and BMC, relative to control group C.*

| Microscopic scores | Mean values - ASC | Mean values - BMC | Mean values - Control | ASC vs. Control | BMC vs. Control |
|---|---|---|---|---|---|
| Morphology | 2 | 3 | 0.33 | 0.01* | 0.002* |
| Matrix staining | 1.67 | 2.83 | 0 | 0.004* | 0.0003* |
| Surface regularity | 1.50 | 2.17 | 0 | 0.001* | 0.01* |
| Structural integrity | 1.17 | 1.67 | 0 | 0.001* | 0.001* |
| Bonding of neo-cartilage with the old | 1.50 | 0.67 | 0 | 0.03* | 0.17 |
| Thickness of neo-cartilage | 1.50 | 1.17 | 0.33 | 0.07 | 0.11 |
| Chondrocyte clustering | 1 | 1.50 | 0 | 0.01* | 0.01* |
| Hypocellularity | 2 | 1.83 | 0 | 0.003* | 0.002* |
| Cell distribution | 1.33 | 2.17 | 0 | 0.001* | 0.001* |
| Degenerative changes in native surrounding cartilage | 2.33 | 3 | 2.33 | 1.00 | 0.18 |
| Tidemark continuity | 1 | 0.17 | 0 | 0.01* | 0.36 |
| **Total microscopic score - cartilage** | **18.83** | **22.17** | **4** | 0.0003* | 0.0002* |
| Subchondral bone reconstruction | 2 | 2 | 1.67 | 0.47 | 0.52 |
| Bonding to adjacent bone | 1.83 | 2.67 | 1 | 0.34 | 0.09 |
| Subchondral bone morphology | 2.67 | 3.17 | 0.33 | 0.004* | 0.001* |
| **Total microscopic score - bone** | **7.50** | **9.17** | **4** | 0.047* | 0.01* |

*p<0.05 statistically significant

| Sheep no. | Group A – ASC | Sheep no. | Group B – BMC | Sheep no. | Group C - Control |
|---|---|---|---|---|---|
| 1 | 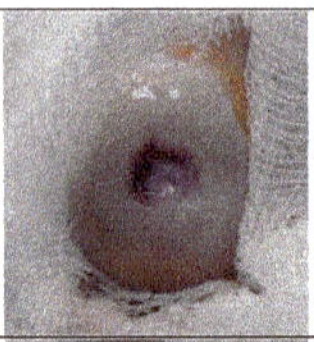 | 7 | 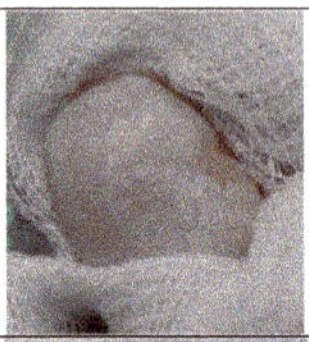 | 13 | 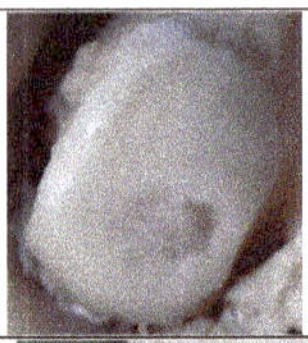 |
| 2 | 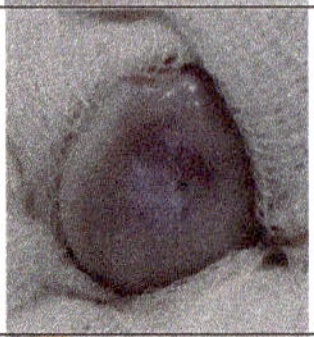 | 8 | 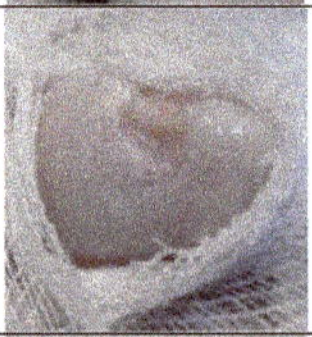 | 14 | 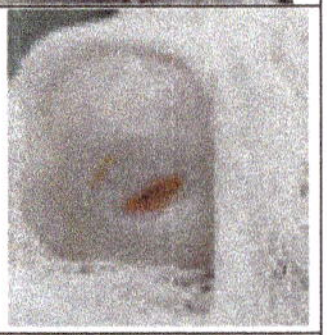 |
| 3 | 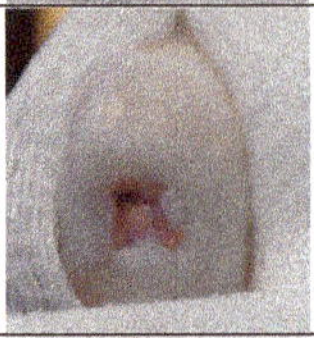 | 9 | 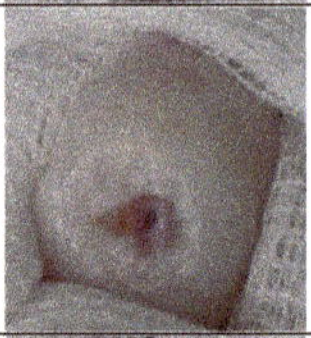 | 15 | 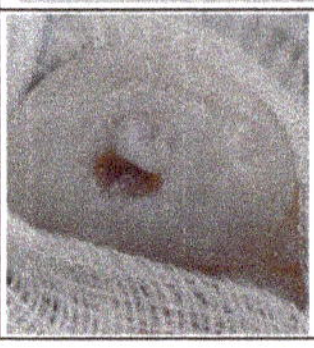 |
| 4 | 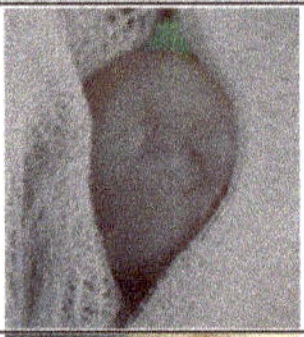 | 10 | 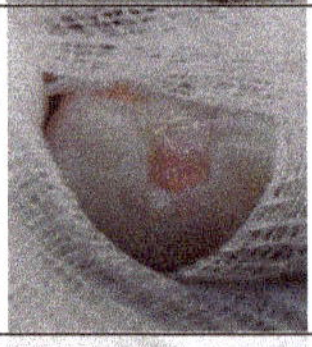 | | |
| 5 | 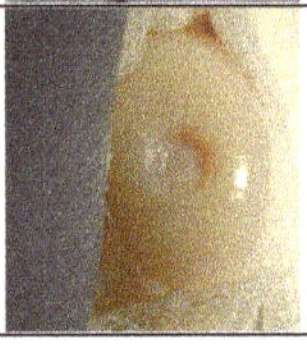 | 11 | 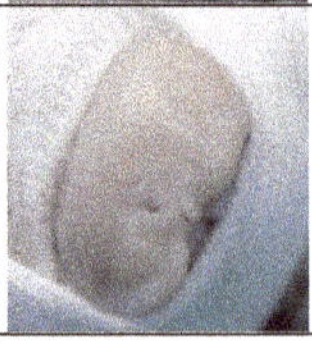 | | |
| 6 | 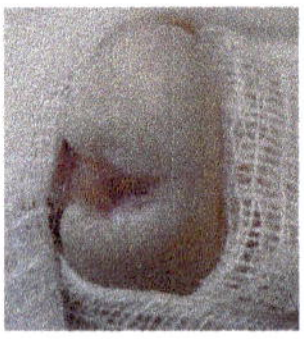 | 12 | 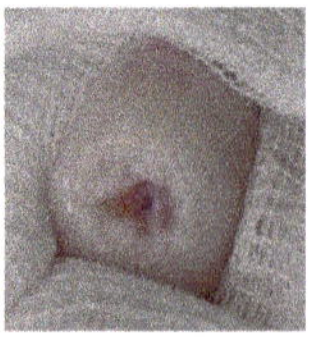 | | |

*Fig. 7.17.*
*Comparative macroscopic images between groups.*

| Sheep no. | Group A – ASC | | | |
|---|---|---|---|---|
| 1 | | | | |
| 2 | | | | |
| 3 | | | | |
| 4 | | | | |
| 5 | | | | |
| 6 | | | | |
| | Group B - BMC | | | |
| 7 | | | | |
| 8 | | | | |
| 9 | | | | |

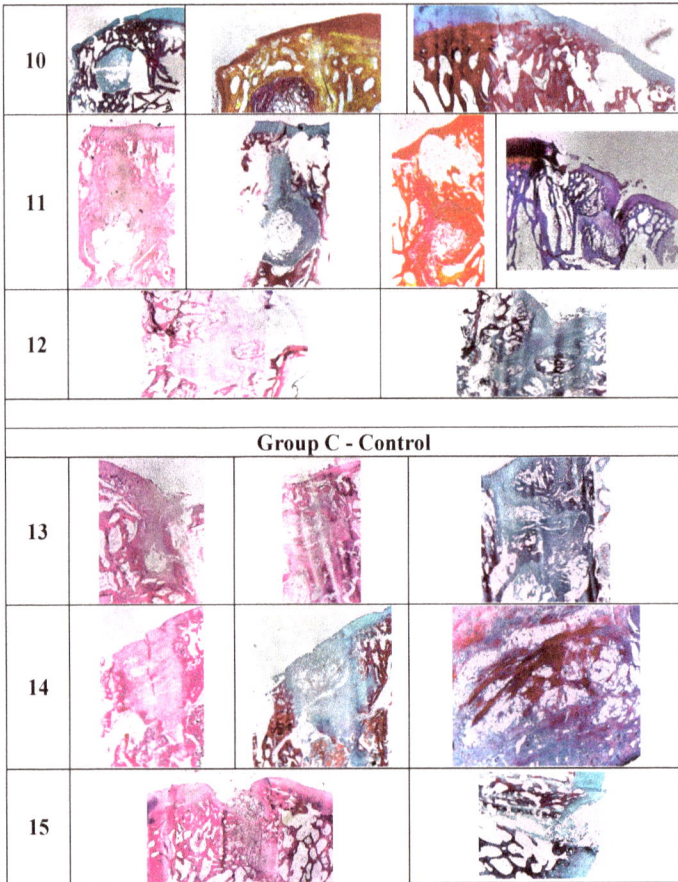

**Fig. 7.18.** *Comparative microscopic images between groups. Staining: Hematoxylin-eosin, Trichrome Masson, Sirius Red, Alcian Blue and Safranin O.*

In Figure 7.17, macroscopic images of lesion healing in the three groups are presented. In Figure 7.18., some histological images with different colorations are compared between treatment and control groups.

It is noteworthy the tendency to widen the bone defect either by its deepening or by the occurrence of some subchondral bone pseudocysts in a few cases (Figure 7.19). It is also noted the occurrence of calcifications in the medullary space (Figure 7.20).

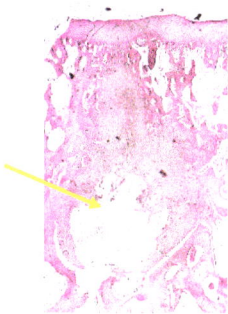

*Fig. 7.19. Subchondral cyst.*          *Fig. 7.20. Focal calcifications.*

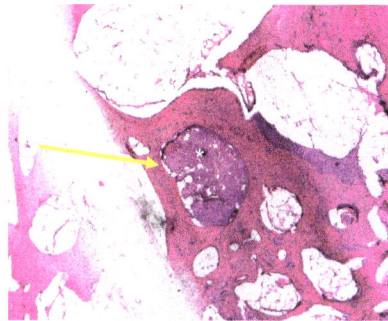

## 7.4.2    Immunohistochemical evaluation

The best images obtained by the immunohistochemical visualization technique of collagen type I and II are shown in Figure 7.2.1a-e. The predominance of type II collagen is noticeable, which signifies the marked tendency of restoring hyaline-type cartilage.

| IHC Image | Comments |
|---|---|
|  Hyaline cartilage / Bone | a. Type I collagen: normally absent in healthy hyaline cartilage. |
|  | b. Type I collagen (orange arrow): Appears around the scar lesion that fills the defect (blue arrow). It is also highly expressed in periocondrum (yellow arrow) and absent in the remaining hyaline cartilage (red arrow). |

| 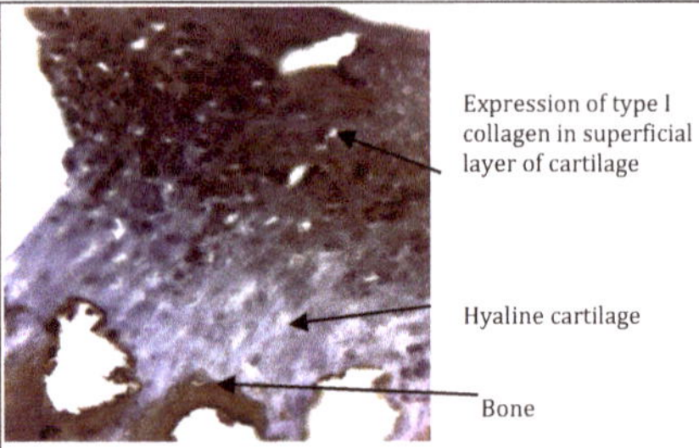 | c. Type I collagen expressed in superficial layer. This fact proves the fibrocartilaginous nature of it. The hyaline cartilage underneath, in contact with the bone, probably comes from maturation of fibrocartilage. |
| 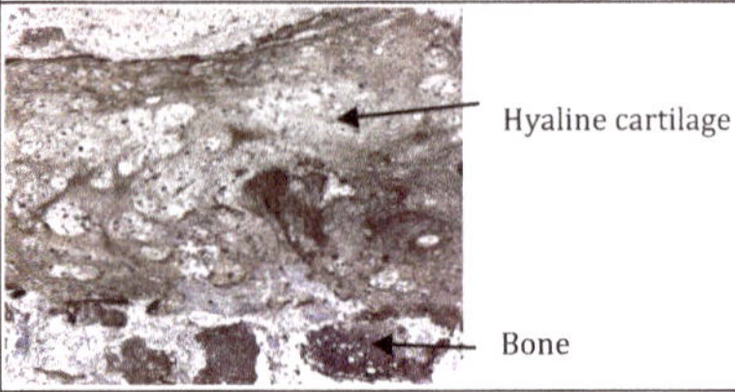 | d. Type II collagen: is expressed in normal hyaline cartilage as a lax network disposed on the force lines of tissue, around chondroplasts. |
| 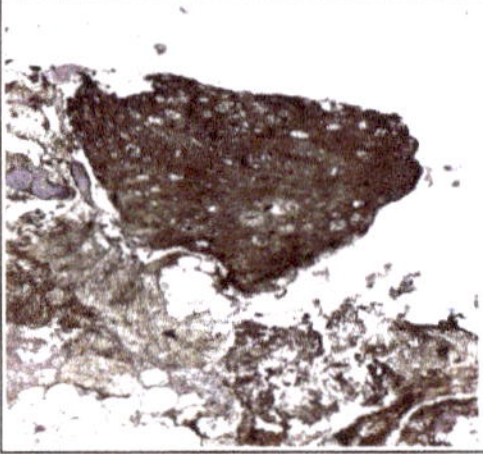 | e. Type II collagen: overexpressed in fibrocartilage area. |

*Fig. 7.21 a-e. Comparative microscopic images between groups.*

### 7.4.3    Imaging evaluation

### 7.4.3.1 Radiological investigations

The radiological investigation is useful for highlighting subcutaneous bone formation. In the present cases, it was observed incomplete restoration of it, in some cases being replaced by scar fibrous tissue (Figure 7.22).

**Group A – ASC**          **Group B - BMC**

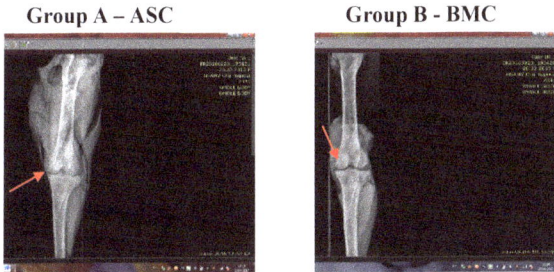

***Fig. 7.22****. Radiologic image of osteochondral defect at 7 months.*

### 7.4.3.2 Computed tomography (CT)

The major concern of the CT examination is the evaluation of subchondral bone recovery (Figure 7.23). However, with insufficient evidence to perform a statistical study, the obtained images are only useful for the qualitative description of the bone regeneration process. Thus, we can observe the tendency of centripetal bone growth, with the formation of an almost complete subcondral bridge, but in some cases we can observe the deepening of the lesion towards the deep area of the defect, in the form of an extension of the defect created or of a pseudocyst, which is also very well observed on the microscopic sections.

### 7.4.3.3 Magnetic resonance imaging (MRI)

The MR imaging obtained, even if it does not allow us to carry out a statistical study, shows the degree of filling of the osteochondral defect and the localization of the newly formed tissue, as can be seen in Figure 7.24, from the periphery to the center of the lesion. The demarcation line of the adjacent cartilage is also partially observed. In some areas, however, there is a fusion with the adjacent tissue. We note the good adhesion of the cartilage layer to the subchondral bone, without distinguishing a net separation line, and the presence of the basal lamina is also observed. We can also see the presence of pseudocysts in the subchondral bone. We do not see the presence of the intra-articular inflammatory edema.

Materials Research Forum LLC
https://doi.org/10.21741/9781644900536

| Group A - ASC | Group B - BMC |
|---|---|

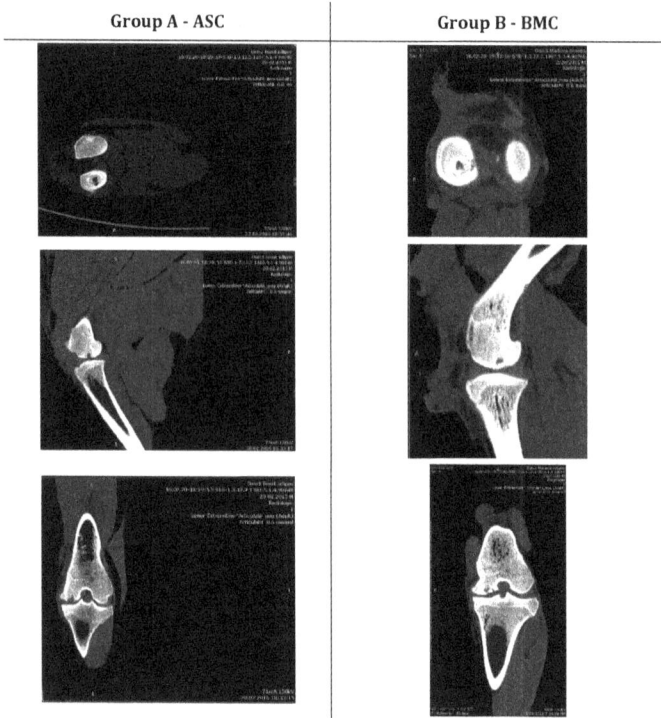

**Fig. 7.23.** *CT scan images of osteochondral defect at 7 months.*

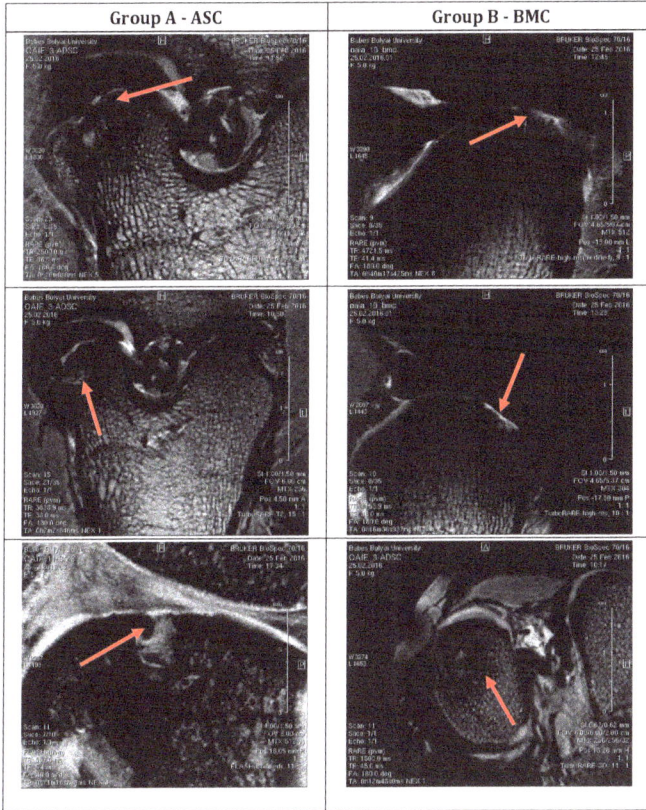

***Fig. 7.24.*** *MRI images after the interventions of osteochondral repair.*

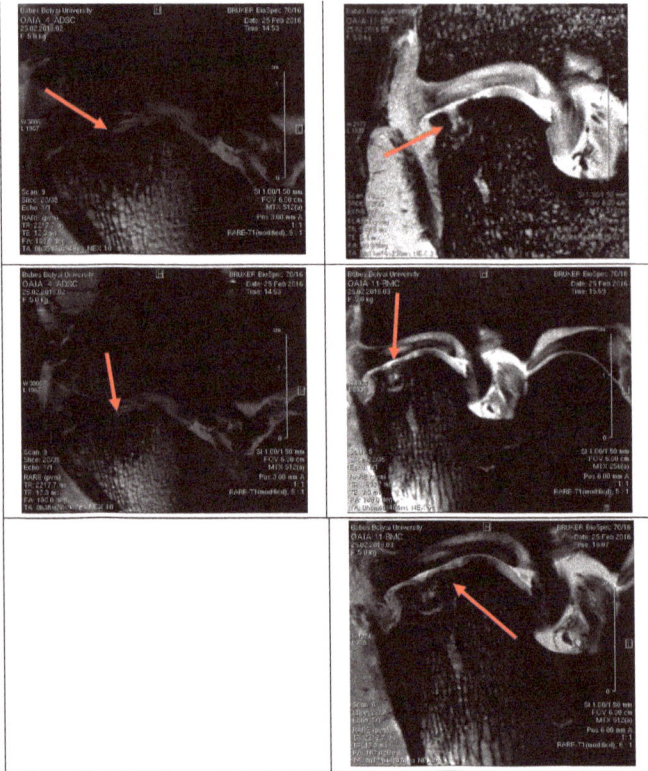

***Fig. 7.24.** MRI images after the interventions of osteochondral repair.*

The surface of the cartilage seems intact, without delamination or fibrillary lesions. The structure looks homogeneous, with a discreet signal alteration on certain sections, reduced in size. The presence of a marginal bone or condral osteophyte is not observed. The subcondral bone shows a slightly hyper signal in some sections, but not bone marrow edema.

## 7.5    Discussions

Although the mean of the Wayne macroscopic score in the treatment groups was higher than in the control group (10.33 vs 8), the difference was not statistically significant due to the lower number of subjects in the control group (3 vs. 6 in the treatment).

Macroscopically, there are no significant differences between groups, but most of the results of the microscopic evaluation are statistically significantly better in subjects treated with multipotent cells. Even though the thickness of the neocartilage is not significantly different between groups, it is important to distinguish the structural quality of the cartilage, as evidenced by morphology, structural integrity and cellular distribution. Also, an important parameter is the integrity of the tidemark, which practically means the adhesion of the cartilaginous layer to the underlying bone, where fat cells appear to be able to give a better result than BMC.

Regarding bone regeneration, there are no quantitative differences between groups, instead, the quality of bone tissue, given by its morphology, is better in the BMC-treated group. Differences are statistically significant in their favor. Focal calcifications found in the matrix could be mineral debris from the inorganic phase of the original biphasic implant, which could not be extracted by decalcification. Our experiment succeeds to prove that the new type of collagen I/ III implant, used for the first time in an animal experiment with multipotent mesenchymal cells, gives very good results in osteochondral regeneration, is biocompatible and stimulates local restoration, without causing important side effects.

Overall, it can be concluded that BMC mesenchymal cells gives the best results in osteochondral regeneration. However, it can be said that cells derived from adipose tissue ASC yields comparable results to the first one. The difference in the presence of these cells from the simple, already classical treatment is obvious, only by applying the collagen membrane and/or performing subcondral microfractures for the local stimulation of mesenchymal cells found in the bone marrow.

Modern imaging, CT and MRI equipment allow us to study the efficacy of treatments without sacrificing animal subjects. The images shown above show the possibilities they offer us in assessing both bone and cartilage tissue. The high-powered MRI device, such as the one used in this study of 7 Tesla, provides high quality images, on which statistical studies by interpreting specific scores, such as the MOCART score for assessing cartilage healing can be performed.

In our case, even if we did not have the possibility to conduct a statistical study, the MRI images allowed us to perform a descriptive study of osteochondral regeneration, providing details of great finesse in addition to the histological study. Thus, the formed subcondral pseudocysts and the relative deepening of the cartilaginous defect could be visualized, as described in histological microscopic images. Further, the researches carried out on the virtual model of the joint with the osteochondral defect, also gives us the explanation of the occurrence of this phenomenon, which will be presented in the next chapter.

Also, for technical reasons, the immunohistochemical study was not carried out on the whole group of subjects in order to be able to statistically analyze the data and draw pertinent conclusions. However, the images that have been made are eloquent to highlight the good result of condral repair with collagen scaffold and stem cells. The images shown in Figure 7.21 show the collagen I/III resorption from the implant and replacement with type II collagen normally found in the cartilaginous hyaline tissue.

At the moment, the issue of the need for cell intake to stimulate condrale healing is being raised by many researchers worldwide. Opinions are divided, literature presents pros and cons of this fact[156,161,169,170,174,225,226]. Another problem that arises is the efficacy of different types of cells that can be used: chondrocytes, mesenchymal stem cells obtained from bone marrow, adipose tissue, circulating blood, synovial membranes, or even umbilical cord blood stem cells. Some studies state the superiority of bone marrow stem cells compared to cells derived from adipose tissue or other sources, also taking into account their higher concentration in the bone marrow[159,160,227-231].

However, the cells in the fatty tissue ASC may present other advantages, at least theoretical: relatively easy harvesting, availability in larger quantities, relatively easier handling and not having the risk of clotting as in the case of BMC. The results obtained by us in this experiment give us reason to believe in the future possibilities of treatment with these AUC cells. The results of the study are encouraging, showing the immense possibilities of research that this field offers, but it is necessary to continue with larger experimental studies of therapeutic methods on animal models and to assess their wider applicability on a larger scale. It is necessary to find simple, reproducible methods with minimal morbidity and risks, cheaper, but with better results and reduced complications.

Comparative clinical trials require homogeneity of patient selection/lesion/ treatment methods, conducting them in a multicenter randomized manner and longer-term follow-up of outcomes. Clear indications for each procedure should be specified by establishing protocols and guidelines. We must have the consensus of all those with concerns in this area to streamline procedures and results.

## 7.6    Conclusions

- Multidisciplinary team research is the premise of achieving the best results for the development of complex therapies in osteoarticular disorders

- The sheep is a very good animal model for therapeutic experiments on the locomotor apparatus, in particular on the treatment of osteochondral lesions, which can go further for studying degenerative joint diseases (osteoarthritis)

- Methods for bone marrow stem cell extraction from the iliac crest as well as adipose tissue are also feasible in sheep, resulting in multipotent mesenchymal cell populations with osteochondral regeneration potential

- Regeneration of hyaline cartilaginous tissue is possible with tissue of the same quality, not just fibrocartilage

- Procedures involving repair in one time, without cell manipulation and laboratory processing are the most effective in terms of the complexity of the therapeutic act but also from an economic point of view

- Mesenchymal stem cells offer real possibilities for treatment of focal cartilage defects, with or without the involvement of the underlying bone

- Mesenchymal stem cells from BMC and ASC clearly demonstrate their superiority to local mesenchymal cells in the subchondral layer

- For better efficacy, it is necessary to use implants that provide support for these cells, but also to restore the architecture of collagen fibers from bone and cartilage

- The collagen implant tested for the first time meets the characteristics required to be used as a support for osteochondral regeneration, providing very good histological, imagistic and clinical results

- BMC cells provide better results in the regeneration of bone and cartilage than ASC cells, but these good results should be also considered due to their relative ease in harvesting and processing them. Their activation with autologous concentrated platelet-rich plasma (PRP) increases their repair potential.

- The results of the study are encouraging, showing the immense possibilities of research that this field offers, but it is necessary to continue with larger experimental studies on larger animal models of therapeutic methods and to assess their wider applicability on clinical scale

- The use of high-performance imaging techniques to evaluate results, such as MRI, CT, should be encouraged, thereby avoiding the sacrifice of animals

- It is necessary to find simple, reproducible methods with minimal morbidity and risks, cheaper, but with better results and reduced complications.

## 8. Study 4. Static analysis with finite elements on the sheep knee treated by osteochondral reconstruction

**The objective** of this research was to develop mathematical models through FEA (Finite Element Analysis), with a high degree of fidelity in representing the real anatomical forms of the femur, tibia and cartilage. The study evaluates the effect of static load stresses by comparing the distribution of stresses in cartilage, cortical and trabecular bone in the treated area of the defect as a result of the in vivo tests on sheep presented in the previous chapter.

**The novelty** of this study is given by the use of FEA analyses on bone structures and cartilage, 7 months postoperatively, after application of the 3 distinct treatments (treatment only with type I/III collagen implants, collagen and stem cells from adipose tissue, collagen and stem cells from the bone marrow). The treated area was located in the left medial femoral condyle cartilage, the central region of maximum convexity. Unlike other studies that analyzed through FEA methods the load-bearing effect on the osteoarticular system[232-235], the current study also describes the effects of possible trauma that can occur on bone or cartilage tissue postoperatively by applying a high pressure of 0.76 MPa. At the same time, the stress distribution and deformation in bone tissue is analyzed with a pressure considered normal for sheep (0.38 MPa).

### 8.1 Introduction

Reconstruction of the skeleton is extremely difficult even for the most experienced surgeons, and some of the critical factors contributing to the complexity of the operative act include: anatomy, the presence of vital structures adjacent to the affected area, the uniqueness of each defect and the chances of postoperative infection. Along with the development of imaging techniques, computed tomography (CT) or magnetic resonance imaging (MRI) has become possible. Using this information, it is possible to reconstruct the *3D* models of bone tissue in the area of interest, after which preoperative customized implants can be made using additive technologies (Additive Manufacturing)[236].

In recent years, various static and dynamic simulations with finite elements have been developed to optimize CAD models of medical implants, encouraging practically the design and manufacture of innovative, customized medical implants in orthopedics[237-239]. To evaluate the stress distribution of bone tissue induced under experimental conditions, many specialists have used finite element analysis[232-235]. Based on these studies, it was hypothesized that FEA studies are an appropriate means of assessing the restoration of cartilage or bone tissue[236,240]. These FEA simulations prevent the destruction of specimens and offer the possibility of testing different materials or conditions. FEA studies have a fundamental importance in acquiring knowledge about the mechanism by which these elements functionally relate (femur - articular cartilage - tibia) and lead to optimization of clinical outcomes.

The purpose of static simulations with finite element is to analyze the mechanical behavior of the knee (femur, tibia and cartilage), under load with two types of pressures. The first scenario is built based on a pressure considered normal for sheep of 0.38 MPa. The second scenario applies a load of 0.76 MPa, which can cause lesions or trauma to bone or cartilage tissues.

Cartilage lesions may result from joint injuries due to fall, direct blows or sports activities, fractures involving the articular surface or fractures due to which a bio-mechanical deficiency occurs in the joint. This phenomenon was analyzed in Scenario II.

For the FEA analysis, the Creo Parametric software was used, with which various simulations were performed to replicate the bone tissue stress after loading, 7 months postoperatively. These simulations have reproduced the biomechanical phenomena that can occur postoperatively in sheep after applying the three types of treatments mentioned.

## 8.2    Evaluation of CT images

### 8.2.1    Methodology

The main stages developed in this study were:

1. Importing and reconstructing *3D* cortical and trabecular bone of the femur and tibia using the MIMICS software; Saving them as STL files; Each case has the following overall *3D* reconstructions: femoral cortical bone, femoral trabecular bone, femoral cartilage, joint fluid, tibial cortical bone, tibial trabecular bone, and tibial cartilage;

2. Verification and correction of errors in *3D* models (STL format) for each bone tissue;

3. Creo Parametric redesign of all *3D* models for each case:

> *Case 0* - represents the healthy group, without surgery, considered as control group (preoperative),

> *Case 1* - represents the sheep group treated only with the collagen implant,

> *Case 2* - represents the group of sheep treated with collagen implant and stem cells from the bone marrow,

> *Case 3* - represents the group of sheep treated with collagen implant and stem cells from adipose tissue;

4. Designing in the Creo Parametric software the joint fluid, the cartilage of the femur and tibia;

5. Exporting from Creo Parametric each of the (Case) in .stp format to the SolidWorks software to optimize the .stp format;

6. Exporting from SolidWorks of each set (Case) in .STEP format defined by AP203 to the ANSYS software;

7. Elaboration of the biological conditions of bone anchoring, ANSYS software;

8. Definition of physical-mechanical characteristics of each tissue, ANSYS software;

9. Scenario I, static FEA simulations by applying a constant and uniform pressure of 0.38 MPa per case, ANSYS software;

10. Scenario II, static FEA simulations by applying a constant and uniform pressure of 0.76 MPa per case, ANSYS software;

11. Interpretation of results on distribution of stresses and total deformities in bone tissue and cartilage.

### 8.2.2    Results of CT imaging

In the present study, CBCT images (Cone Beam Computed Tomography) of 3-year-old sheep, clinically healthy, were used. CBCT images were taken with a Siemens scanner (*Scope, Germany*, equipped with the Syngo CT VC28 software) 7 months postoperatively, after applying the three types of treatments mentioned. The following parameters were used: 130 kV, 7 mA, 200 sections maximum, 0.14 mm layer thickness, 13.5 cm x 22.5 cm visual field, and 12 s exposure time.

*Case 1 - collagen treatment only (**Annex 8.1**)*

- explaining the phenomenon by which the bone retracts in depth up to a maximum distance of 8-9 mm from the cartilage area (**Annex 8.1**, Figure 5, the left upper image); However, in the axial sections (Figure 8.1, Figure 5, top right image) the diameter of the defect is reduced from an initial diameter of 8 mm to approx. 6.3 mm, resulting bone and cartilage development near the sidewalls of the defect (**see Annex 8.1**).

- the cortical bone is not restored, only partially, in the area adjacent to the defect (**Annex 8.1,** Figure 3).

*Case 2 - treatment with collagen and bone marrow stem cells (**Annex 8.2**)*

- the treated area has relatively uniformly developed bone tissue, which is identified using the "Measure Density inside Ellipse" function of the MIMICS software. With this tool, an ellipse marked in all sections of Annex 8.1, 8.2 and 8.3 was marked. Inside the ellipse, bone density was measured, which in this case has a value between 1650 and 2010 on the contrast scale (see **Annex 8.2**). This range corresponds to adult bone tissue. The homogeneity of the newly formed bone can be seen in **Annex 8.2,** Fig. 1, 2 and 4 where the variation of the contrast scale within the ellipse is limited (see the "Std. Dev." values in Figures 1, 2 and 4).

- both cortical, trabecular bone and cartilage were developed in almost the entire area treated (Figures 1, 2, 3 and 5), with only one exception in section 66.26 (**Annex 8.2**, Figure 4).

*Case 3 - treatment with collagen and stem cells from adipose tissue (**Annex 8.3**)*

- CT imaging shows that the bone area only partially developed in the medial area of the treated region after 6 months postoperatively, and in Figure 2 of **Annex 8.3** it can be seen a portion there is new bone tissue (6.4 mm deep and 2.8 mm width). Practically, with this treatment, the phenomenon of retraction identified in Case 1 is lower in both depth and width.

- the new trabecular bone tissue developed mainly in the area of the lateral walls of the incision, and the cortical bone has activated its growth, partially embracing the incision (**Annex 8.3**, Figure 3, Section 70.17).

### 8.2.3   General stage conclusions

The new bone and cartilaginous tissue developed primarily on the sidewalls of the defect. The initial diameter of the incision was 8 mm and the depth was 4-6 mm. In Case 1, the trabecular bone developed only 2-3 mm, restricting the incision diameter to 5-6 mm. In Case 2, both trabecular and cortical tissue and cartilage tissue were stimulated and developed homogeneously throughout the treated area. In Case 3, the trabecular and cortical bone developed partially. It has been noticed that in the apex of the incision there is a phenomenon of bone retraction, with the formation of a pseudo-cystic area, which corresponds to the findings of the histological examination presented in the previous chapter. This phenomenon is more pronounced in Case 1, where the lack of trabecular bone goes up to a depth of 8-9 mm from cartilage.

This valuable information superimposed by the histological study supports the idea that in areas where no new bone and cartilage tissue has developed, there is fibrocartilage that is mainly composed of type 1 collagen.

### 8.3   *3D reconstructions of the knee*

In order to reconstruct *3D* models, CBCT images were imported into the MIMICS software (*Materialise, Belgium*), where masks for the cortical and trabecular bone of the femur were obtained, as well as for the cortical and trabecular bone of the tibia. These *3D* masks required complex processing operations to make the bone tissues as accurate as possible. Finally, they were exported in .STL format to the Creo Parametric software (*Parametric Technology Corporation, USA*). In this software, sheep cartilage was designed according to some studies in the field[241] as follows: left femur cartilage has a variable thickness of 0.9-1.1 mm and the left tibial cartilage also has a variable thickness of 1.1-1.3 mm. These *3D* cartilage models correspond to Case 0 (control group, preoperative, Figure 8.1) and Case 2, where histological analysis revealed that both the cartilage and the cortical bone are recovered.

Case 2 illustrated in Fig. 2b and Fig. 6 of **Annex 8.2**, however, presents a slight deviation from the original surface. In Case 1, the treated area does not have cartilage surfaces, only collagen (Figure 8.2a). Case 3 has cartilage surfaces but is interrupted by collagen (Figure 8.2c). Following closely the bone defect illustrated in CT images (see **Annex 8.1**, **8.2** and **8.3**), these areas were designed as accurate as possible.

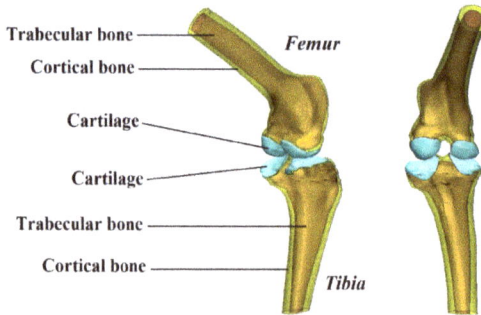

*Fig. 8.1.* *3D bone and cartilage reconstruction of Case 0 – healthy.*

In Figure 8.1 we can see the 7 analyzed and projected landmarks that make up a femoral-tibial ensemble (including synovial fluid). Synovial fluid of the joint was designed with a variable thickness (minimum 0.03 mm) in order not to influence the results, according to studies[232,242].

In order to give a true picture of the real cases treated, each procedure was repeated, focusing on the area where the surgical intervention occurred (Figure 8.2 - left lateral femoral cartilage, central region). Thus, the four assemblies were formed as follows: Case 0 - represents the healthy group also considered as the control group (preoperative); Case 1 - is the group of sheep treated only with collagen; Case 2 - is the group treated with collagen and stem cells from the bone marrow; Case 3 - is the group treated with collagen and stem cells from adipose tissue.

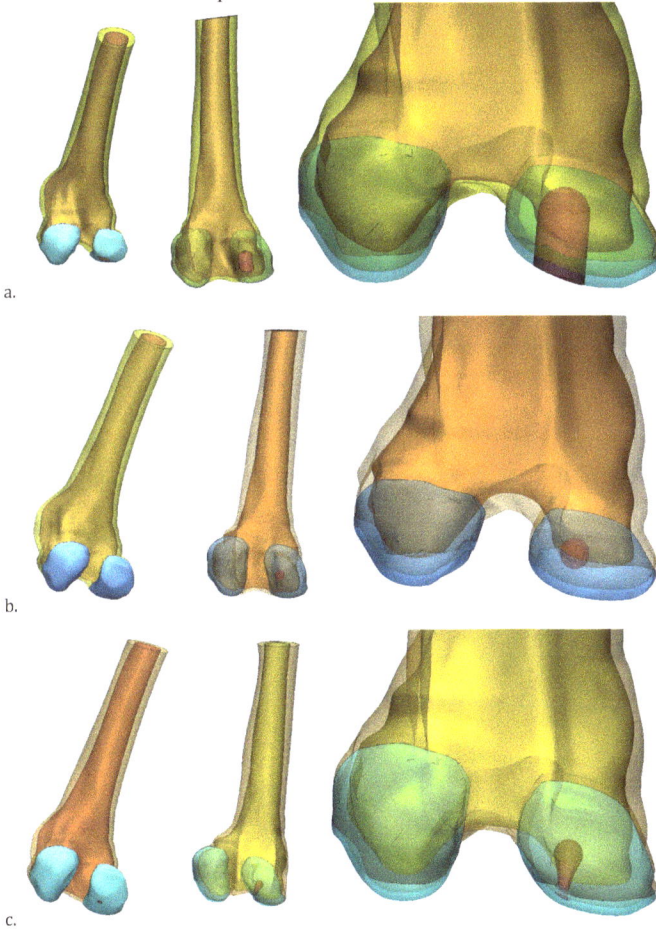

***Fig. 8.2.*** *The defect zone treated and 3D reconstructed: a) Case 1; b) Case 2; c) Case 3.*

### 8.3.1   FEA static analysis conditions

Finite Element Analysis (FEA) is based on the concept of building complex models using simple elements or the division of complex objects into small, easy-to-handle pieces[236]. The graphical interface provided by the Creo software allows structures to be analyzed by configuring inputs close to real cases, followed by the interpretation of von Mises criteria. Knowledge of the stress state and deformations in the structure are indispensable in understanding the post-operative phenomena after surgical treatment, and the distribution of forces is relevant in analyzing manipulative critical areas[241].

Using the Creo Parametric software, the femoral-tibial ensemble was statically analyzed for compression. To calculate the pressure that exists in the osteoarticular system of the knee, the following equation was used[232]:

$$p = \frac{F}{2(\pi ab)}$$

Where: p - the pressure expressed in [N/mm²]; F - the force with which the femur is loaded [N]; a and b - ellipse diameters for a bone section [mm].

In figure 8.3 are the dimensions of the cortical bone in axial view rendered. The force acting on the knee joint of a sheep is 2.1-2.25 times the weight of the animal, according to studies[232,233,243]. In our case, the sheep's weight was about 50 kg (2-3 years old), resulting in a force of 1120 N. By inserting this information into equation (1) results a pressure of 0.38 MPa, taking into account the diameters of the cortical bone section (Figure 8.3a).

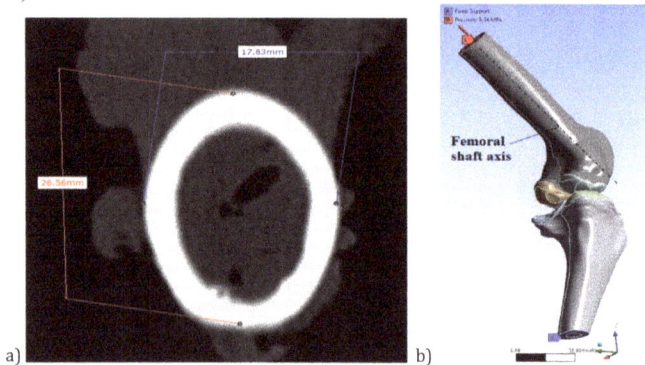

***Fig. 8.3. a)*** *Maximal dimensions of cortical bone, axial view of left femur;*
***b)*** *FEA conditions in ANSYS software: A – constraint zone; B – pressure zone and direction.*

*Table 8.1. Physical and mechanical characteristics of different tissues in the knee.*

| Material | Young modulus [MPa] | Poisson coefficient | Density [g/cm³] | Compression resistance [MPa] | References |
|---|---|---|---|---|---|
| Trabecular bone | 452 | 0.3 | 0.41 | sub 5 | 232,244-246 |
| Cortical bone | 16160 | 0.33 | 2.45 | 100-147 | 232,247,248 |
| Cartilage | 0.8 | 0.4 | 1,1 | 1-20* | 232,249-251 |
| Type 1 collagen | 5000 | 0.3 | 1.32 | 50-100 | 252-256 |
| Synovial fluid | 1 | 0.499 | 1 | - | 232,242 |

* Tension that causes trauma in the cartilage[257]

This constant and uniform pressure was applied in Scenario I, and it may represent a normal situation in which the osteoarticular system is subjected to daily movements. In Scenario II, a high and uniform 0.76 MPa pressure was applied to the upper plane of the femur, which could cause trauma or lesion. The pressure conditions imagined in Scenario II are intended to mimic real situations that may occur in the case of bone trauma. In each scenario, the pressure direction was parallel to the femoral diaphyseal axis (Figure 8.3b). In order to accurately reproduce a real static situation, the ensemble was constrained in the lower plane of the tibia in all directions conform[235] (see Figure 8.3b). At the same time, 17-21 contacts were set between the *3D* models, depending on collagen intercalations with bone and cartilage tissues.

Knowing that each *3D* model designed and illustrated in Figure 8.4 has specific physical and mechanical characteristics, Table 8.1 details the properties of bone, cartilage tissue and synovial fluid, respectively. These physico-mechanical characteristics were attributed to the 3D models of the femoral-tibial ensemble. It is also assumed that they are homogeneous, isotropic and have a linear elastic behavior. Recent studies have shown that the Poisson coefficient of cartilage is less than 0.4, sometimes approaching zero [232,250,251]. In the past, cartilage was supposed to be incompressible, and therefore the Poisson coefficient had a high value of 0.49. Synovial fluid of the joint was considered as an impermeable membrane with a variable thickness (minimum 0.03 mm) and a Poisson coefficient of 0.499 (the maximum admitted in the FEA analyses).

## 8.3.2    FEA static simulation results

Initially, the classic FEA simulation method in the ANSYS software was carried out, but major deficiencies were found both in the discretization of the models obtained from

.STL files and later in the transfer of the stresses between the models, generating the phenomenon of „geometrical concentrators of tension".

It has been experimentally found that sometimes in narrow (local) areas the stresses are much higher, and this phenomenon is known as „stress concentrations". These areas of excessive stress are generally the contact between surfaces of *3D* models analyzed with finite elements, especially when the parts with anatomical forms specific to the medical field are simulated. Stress concentrations occurs in the following cases:

- The *3D* model has notches or sharp edges (local contacts);

- Contact surfaces between 2 parts are discontinuous or have sudden variations.

This phenomenon leads to errors in generating the stress state of the femur-tibia ensemble. In view of these observations, all the models of the femoral-tibial ensemble have been parameterized using CAD methods. To eliminate these stress concentrators that appeared in the *3D* collagen model, this element was re-engineered by insisting on the curvature of the surfaces (for example in Case 2, the *3D* model of collagen is a sphere).

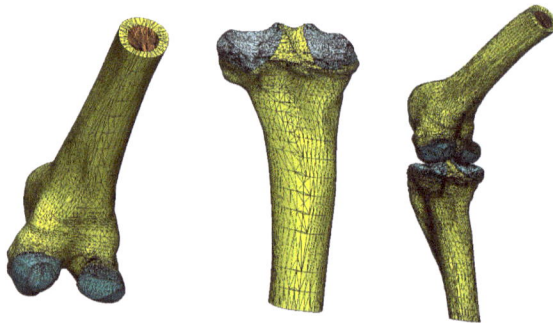

**Fig. 8.4.**  *Discretized 3D models.*

**Table 8.2.** *Discretization of 3D models*

| *3D* **Knee Assembly** | **Nods** | **Finite Elements** |
|:---:|:---:|:---:|
| *Case 0* | 675 947 | 385 300 |
| *Case 1* | 679 417 | 386 356 |
| *Case 2* | 677 913 | 386 279 |
| *Case 3* | 638 965 | 389 932 |

The discretization of each *3D* model was performed in the ANSYS software using the "Mesh" command, which set up fine meshes with the size of an element of 0.5 mm, resulting in the values of the nodes and finite elements detailed in Table 8.2. The nodes are interconnected by finite elements in tetrahedral form (Figure 8.4). Even if a high-performance graphics station was used for FEA analyzes, the use of these fine discretization meshes led to increased static runtime (approximately 2h for each case), but augmented the accuracy of the results.

In order to be able to identify more easily the values of the equivalent stresses von Mises, Fig. 8.5-8 and Fig. 8.10-13 have the same scale of values. On this scale were marked the limit values for compression resistance of trabecular bone (5 MPa), cartilage (1 MPa, 20 MPa), type 1 collagen (50 MPa) and cortical bone (100 MPa), values identified from other studies listed in Table 8.1. Each of the aforementioned figures has a sagittal and coronal overview, as well as a section with stress status in the defect area.

In Tables 8.3 and 8.4 the maximum values of the von Mises equivalent stress and the total deformation recorded for each case in the area of interest (defect) are reproduced after Scenario I and II simulation. These tables show the maximum stresses in the femur for cortical, trabecular bone and cartilage.

*Table 8.3.* *Results of FEA simulations after Scenario I\*, in the defect zone.*

| Distal femur | Maximal von Mises Tensions [MPa] | Total Deformations [mm] |
|---|---|---|
| *Case 0* Cortical bone/ Trabecular bone/ Cartilage | 3 / 0.5 / 0.5 | 1-3 / 2-3 / 0,5-2 |
| *Case 1* Cortical bone/ Trabecular bone/ Cartilage | 5 / 0.5 / 1 | 1-3 / 2-3 / 1-2 |
| *Case 2* Cortical bone/ Trabecular bone/ Cartilage | 10 / 1 / 0.5 | 1-3 / 2-3 / 1-2 |
| *Case 3* Cortical bone/ Trabecular bone/ Cartilage | 10 / 1 / 0.5 | 1-3 / 2-3 / 1-2 |

\*The action of a constant 0.38 MPa pressure is considered normal

***Table 8.4.*** *Results of FEA simulations after Scenario II**, in the defect zone.*

| Distal femur | Maximal von Mises Tensions [MPa] | Total Deformations [mm] |
|---|---|---|
| **Case 0** Cortical bone/ Trabecular bone/ Cartilage | 10 / 0.5 / 3 | 2-4 / 3-4 / 1-3 |
| **Case 1** Cortical bone/ Trabecular bone/ Cartilage | 20 / **3-5** / **3-7** | 2-4 / 3-4 / 1-3 |
| **Case 2** Cortical bone/ Trabecular bone/ Cartilage | 20 / **3-5** / 3 | 2-4 / 3-4 / 1-3 |
| **Case 3** Cortical bone/ Trabecular bone/ Cartilage | 20 / **3-5** / **3-7** | 2-4 / 3-4 / 1-3 |

**The action of a constant 0.76 MPa pressure can cause osseo-cartilaginous lesions

In general, the modality of stress transmission is similar in all cases, with differences only in the area of defects. High stresses were recorded in the lower plane of the tibia because it is the region where the femoral-tibial ensemble is constrained. Here, according to Scenario I, the stress in the cortical bone of the tibia is 10-20 MPa, and in Scenario II increases to 30-50 MPa. Even assuming that an excessive force would be applied postoperatively at 6 months, the stress in the tibia is 50% below the fracture limit by compression of the cortical bone (100 MPa), details in Figures 8.10-13.

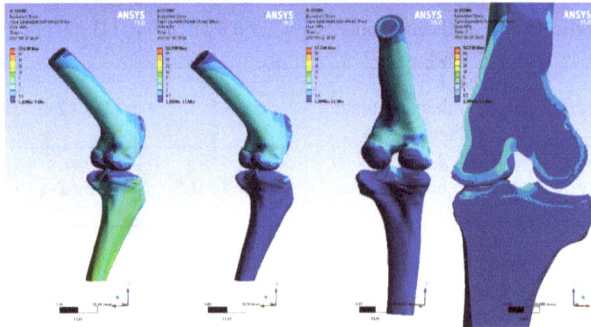

***Fig. 8.5. Case 0***, *distribution of equivalent von Mises tensions in Scenario I.*

In Figure 8.5 the distribution of equivalent von Mises stress may be observed under normal conditions of the healthy femoral-tibial ensemble. In Case 0, the cartilage is loaded below 0.5 MPa, and the cortical bone in the femur takes up the higher stresses and transmits them to the bone tissue of the tibia. Basically, the trabecular bone is not stress loaded, it only transmits it, which is explained by its high elasticity.

After treating the defect only with collagen, applied in Case 1, the state of stress under normal conditions is illustrated in Figure 8.6. It can be seen that this type 1 collagen substitutes both the bone tissue (compact and spongy bone) and the cartilage in the defect area. Thus, it loads with stress between 1-3 MPa which it transmits throughout its volume. Being much more rigid than the spongy bone, the Young modulus being 5000 MPa vs. 0.8 MPa, this collagen reaches the contact also with this tissue, stresses of 1-3 MPa approaching the fracture resistance of the spongy bone (4 MPa).

This aspect can lead to the explanation of the phenomenon whereby the trabecular bone is retracting in the apex of the defect, even under normal conditions *in vivo*. Analysis of CT images supports this hypothesis, finding that at 6 months after surgery, the defect depth increases from 4-6 mm to 8-9 mm. In addition, the collagen in the defect gets in direct contact with the cartilage of the tibia, and thus distributes at their contact stresses of 1 MPa, that can damage the cartilage.

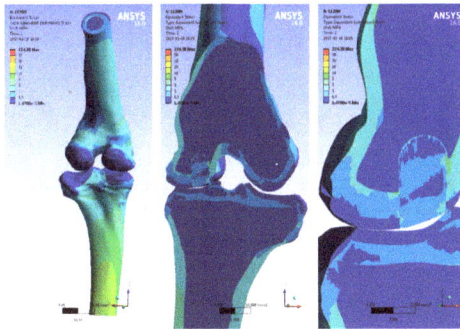

***Fig. 8.6. Case 1**, distribution of equivalent von Mises tensions in Scenario I.*

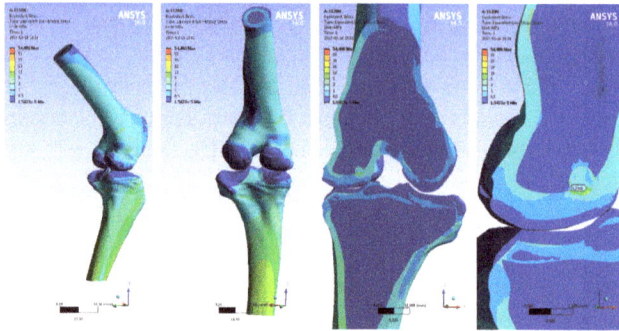

***Fig. 8.7. Case 2***, *distribution of equivalent von Mises tensions in Scenario I.*

The distribution of the von Mises stresses in Case 2 is illustrated in Figure 8.7, where it can be seen that the collagen does not penetrate the cartilage (which has recovered within 6 months) and thus does not end up transmitting stresses to their contact. This behavior is similar to Case 0, specific to healthy bone tissue. Having a low volume, the collagen in the defect area loads with stresses from contact with the cortical bone and distributes it to the trabecular bone a stress that does not affect it (limit to 1 MPa).

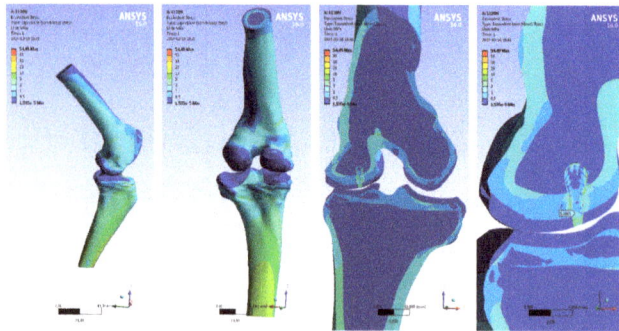

***Fig. 8.8. Case 3***, *distribution of equivalent von Mises tensions in Scenario I.*

Figure 8.8 illustrates the effect of loading with normal pressure on the osteoarticular system of the knee afferent to Case 3. Since cartilage and cortical bone have not fully recovered, there is collagen in contact with these tissues. Thus in this case, collagen loads with stresses and distributes them into adjacent tissues. Compared with Case 1, collagen has a lower volume, which limits the contact surfaces between it and the trabecular bone, respectively the cartilage.

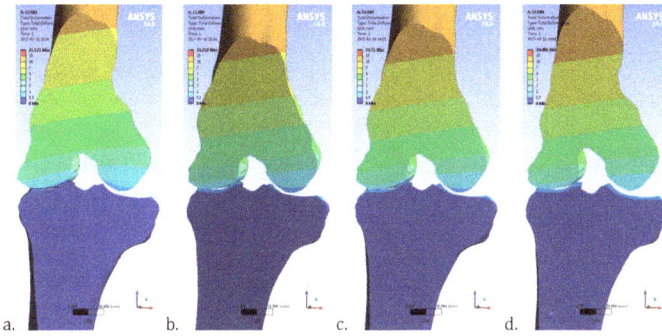

***Fig. 8.9.*** *Total deformations in Scenario I: a)* ***Case 0****; b)* ***Case 1****; c)* ***Case 2****; d)* ***Case 3***

At 6 months postoperatively, following the application of a force of 1120 N, the femoral cartilage of the healthy bone has a total deformation between 0.5 to 1 mm on contact with the tibial cartilage (Case 0, Figure 8.9a). After application of the 3 treatments, the areas where the bone defects were located suffer total deformations of 1-3 mm for the cortical bone, 2-3 mm for the trabecular bone and 1-2 mm for the femoral cartilage (Table 8.3).

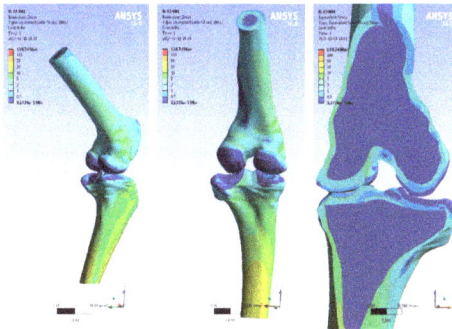

***Fig. 8.10.*** ***Case 0****,* distribution of equivalent von Mises tensions in Scenario II

Figure 8.10 presents the results of application of an excessive static force for sheep (2240 N), having the direction parallel to the axis of the femoral diaphysis. FEA simulations show that femoral cartilage is loaded with a tension between 0.5-3 MPa, the highest level being in the contact area with the tibial cartilage. When the cartilage is loaded, the joint fluid flows from its solid matrix, which reduces the volume of the whole cartilage. Since this tissue is a solid and fluid mixture, the cartilage behaves like a compressible material that absorbs stresses and transmits them totally[251]. Also in this case it has been found that bone tissues distribute the stress and there are no areas where the maximum stresses are above the resistance to compression of the tissue in question.

After application of 0.78 MPa pressure in each treated case, the collagen in the defect area is stress loaded from the cortical bone and transmits them through its whole volume to the trabecular bone and to the lateral walls of the newly formed femoral cartilage (Case 1 - Figure 8.11 and Case 3 - Figure 8.13). The collagen in Case 1 also affects the cartilage of the tibia as it transmits a 3 MPa stress on their contact surfaces (Figure 8.11). Thus, in the newly formed cartilage of the femur there are areas with stresses of 5-7 MPa, which can damage the tissue according to some studies from the literature[257,258]. In addition, collagen transmits 3-5 MPa stresses at the apex of the defect, on contact with the trabecular bone.

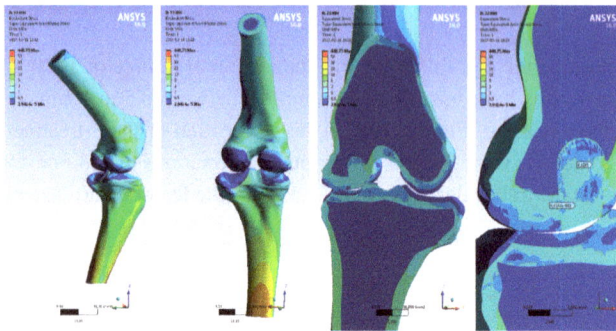

***Fig. 8.11. Case 1***, *distribution of equivalent von Mises tensions in Scenario II.*

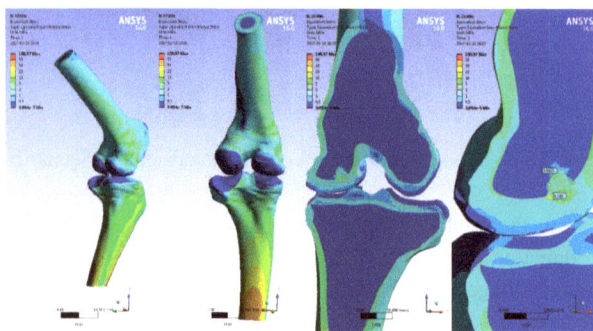

***Fig. 8.12. Case 2***, *distribution of equivalent von Mises tensions in Scenario II.*

The stress state of Case 2 is shown in Figure 8.12, where it is observed that the stress in the femoral cartilage is maximum 3 MPa in the contact area with the cartilage of the tibia. The distribution of stress in cartilage is similar to that of the healthy bone. After applying the treatment in Case 2, the cartilage is restored at 7 months postoperatively, but some complex movements involving high forces (e.g. 2240 N) could affect regions of the restored trabecular bone. The collagen is loaded with stresses from the cortical bone and transmits in the contact area with the trabecular bone stresses between 3-5 MPa. This phenomenon can also be seen in Case 3 of Figure 8.13. Over time, this increased stress in the trabecular bone could lead to its erosion.

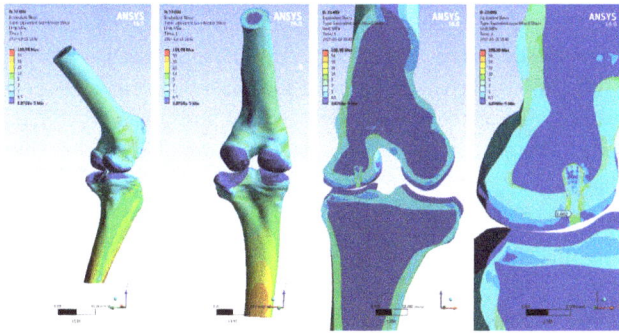

**Fig. 8.13. Case 3**, *distribution of equivalent von Mises tensions in Scenario II.*

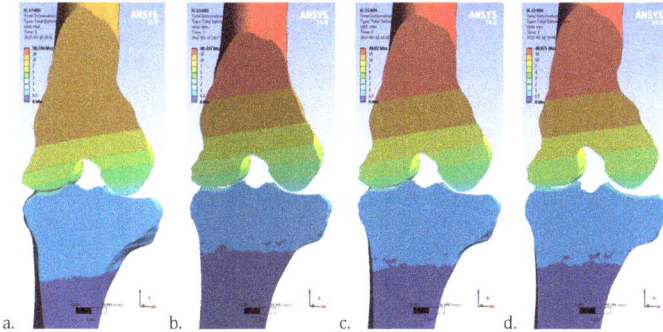

**Fig. 8.14.** *Total deformations in Scenario II: a)* **Case 0**; *b)* **Case 1**; *c)* **Case 2**; *d)* **Case 3.**

At the time of applying excessive force on the osteo-articular system, the collagen in the defect area was in direct contact with the opposing bone tissues, and the total deformations in the defect region increased to 1-3 mm (Figure 8.14). The collagen in the defect, existing in Cases 1 and 3, stretched laterally, producing a concentration of stress on both the sidewalls and the apex of the defect. Total deformation values for each femur tissue are detailed in Table 8.4.

## 8.4    Conclusions

In the simulated scenarios with 1120 N (normal) and 2240 N (excessive) forces, it is demonstrated that there are no stresses the cortical bone of the femur or tibia that is close to its fracture limit, in reality being below 50 MPa.

From FEA simulations on Case 0 - healthy, it turns out that cartilage behaves like a compressible and elastic material that absorbs stresses and distributes them into adjacent bone tissue. This cartilage behavior is similar to that of the periodontal ligament (part of the dental tissues). After applying an excessive static force, the healthy femoral cartilage is loaded with a stress of 3 MPa at the contact with the tibial cartilage. Similar results have been obtained in Case 2, where the cartilage has a stress ranging between 0.5-3 MPa, a stress that is considered normal for cartilage contact.

After applying the treatment in Case 2, the cartilage is restored at 7 months postoperatively, and normal movements distribute stresses into tissue similar to Case 0 - healthy. Complex movements involving high forces could affect regions in the restored trabecular bone.

The existing collagen in the defect at 7 months is stress loaded from the cortical bone and it distributes directly into the other tissues with which it is in contact, reaching to transmit high stresses in both the trabecular bone of the femur and in cartilage. This phenomenon could particularly damage the lateral walls of the newly formed cartilage of the femur, but also the tibial cartilage with which they come into direct contact, visible in Cases 1 and 3 (Figure 8.11 and 8.13). The behavior of collagen to lead stresses is explained by its rigidity compared to elastic tissues (spongy bone and cartilage).

In Cases 1 and 3, after applying a force of 2240 N, the newly formed cartilage in the defect area is stress loaded in some areas reaching at 5-7 MPa. This level of stress can damage the cartilage. The evidences available in the literature suggests that some disorders with cartilage damage are associated with cartilage stress between 1-20 MPa. The existence of collagen in the defect area leads to an increase in deformations both normal situations (1-2 mm) and after applying an excessive pressure (1-3 mm).

When assessing the biomechanical response of cartilage of the animals treated in vivo, FEA studies may help to make clinical decisions about how and what kind of materials can be applied in the treatment of human cases, knowing that without effective intervention, the progressive loss of the affected cartilage may be promoted.

## 9. Study 5. Retrospective observational study on articular cartilage therapies performed in clinical practice

### 9.1 Introduction

The morphological and metabolic particularities of cartilaginous tissue have aroused a lively interest among clinicians and researchers. At the moment, there are very few clinical trials on large groups, on the basis of which objective conclusions can be drawn on the indications and treatment results of articular cartilage lesions[46-49,129].

The incidence of articular cartilage lesions was reported as 20-25% in the general population, of which only 10% are symptomatic. Causes include single or repeated joint injuries, severe sprains, osteochondral fractures, osteochondritis dissecans, rheumatic inflammatory diseases, degenerative diseases, or sometimes they may be iatrogenic. The most commonly affected are the knee joint, followed by the tibio-tarsal joint. Assessment of injuries should take into account several criteria: patient's age, sport activity level, lesion localization, within or outside of the bearing surface, lesion's depth, extension and stability, joint alignment, and the presence of osteoarthritis and its severity[46-48,129].

The clinical symptomatology is nonspecific, the main symptom being pain, mechanical, sometimes inflammatory in nature, occasionally associated with joint blockage. The structure of cartilaginous tissue, consisting of collagen fibers arranged on four layers, cells and ground substance; lack of vascularity and reduced proliferation capacity of chondrocyte cells are responsible for spontaneous healing incapacity, progressive destruction of the entire cartilage and arthrosis[129].

There are currently several therapeutic procedures that attempt to repair cartilage lesions. None of them has proven efficacy, the results being mediocre, both in terms of the quality of the repair tissue, and in terms of improving the symptomatology and joint functionality[231]. The major problem that arises is to establish precise indications of treatment, according to the type and extension of the lesion, age and level of activity of the patient[49].

### 9.2 Work hypothesis

In the light of these considerations, we considered it appropriate to carry out a clinical study on this subject in order to obtain results that could help improve treatment guidelines for articular cartilage injuries. The study also aims to provide data on epidemiology and lesional associations.

In recent years, biological therapeutic methods have been proposed that involve the use of multipotent stem cells and biologic materials for tissue regeneration support, which appear to yield superior results to "classical" methods. In this regard, the hypothesis of this study is that patients receiving such treatments will have at least the same results as those already validated for the same type of injury. Thus, the aim of the study was to evaluate preoperative and postoperative joint functional scores at 6 months for patients undergoing these therapeutic procedures.

### 9.3 Materials and methods

Between 2013-2016 we performed a number of 427 knee and ankle arthroscopies in Orthopedics-Traumatology Clinic, Section II, Cluj-Napoca, Romania. Out of these patients, 197 (46%) had joint cartilage lesions of varying degrees and extensions. Of these, we selected the study group based on the intraoperative findings, which corresponds to Outerbridge III and IV articular cartilage lesions, either isolated or in combination with other pathologies. Superficial lesions that did not require specialized treatment, in 119 cases (60%), were not taken into study. The distribution of cases according to the involved joint and the treatment methods used is presented in Table 9.1.

*Table 9.1. Distribution of cases in groups according to the joint involved and the treatment practiced.*

| JOINT | TREATMENT | NO. OF CASES | TOTAL TREATMENT | TOTAL CHONDRAL LESIONS | TOTAL ARTHRO-SCOPIES |
|---|---|---|---|---|---|
| KNEE | 1. DEBRIDEMENT | 37 | | | |
| | 2. MICROFRACTURES | 24 | | | |
| | 3. OSTEOCHONDRAL TRANSFER | 7 | 75 | 192 | 420 |
| | 4. COLLAGEN MEMBRANE | 7 | | | |
| ANKLE | 1. MICROFRACTURES | 2 | | | |
| | 2. ILIAC CREST BONE MARROW ASPIRATE INJECTION | 1 | 3 | 5 | 7 |

Due to the relatively small number of ankle arthroscopies, it was not possible to carry out a statistical study of these, and are to be presented only descriptively.

We performed the statistical (retrospective observational) study on a group of 75 cases of knee chondropathy with surgical treatment indications. The cases were divided into four groups (1-4) according to the procedure used, as stated in the table above.

The inclusion criteria in this study group were: 16-70 year old patients with Outerbridge grade III or IV cartilage lesions subjected to therapeutic arthroscopy for previously identified lesions (since arthroscopy is currently not performed for the sole purpose of diagnostic). The lesions taken into consideration were focal, bounded by healthy cartilage tissue, even if the joints exhibited signs of incipient arthritic degradation.

The dimensions of the lesions by direct measurement with the probe and the arthroscopic ruler, their number and location (medial or lateral femoral condyle, femoral trochlea, patella and medial or lateral tibial plateau) have been noted.

The therapeutic methods were applied according to the indications of each (described more extensively in the treatment chapter of cartilage lesions). The methods were as follows:

- debridement - regularization with the shaver or scissor of the edges of the lesions, excision, of partially mobile or free cartilage fragments, of intra- and peri-lesional fibrous tissues to healthy tissue (Figure 9.1.a);

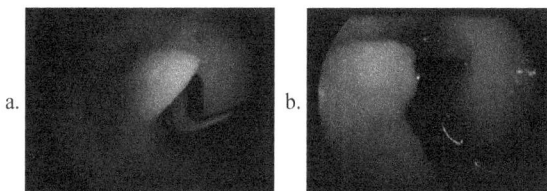

*Fig. 9.1.* a. *Lesional debridement; b. Microfractures.*

- *Microfractures* - belongs to the group of bone marrow stimulation techniques. From a technical point of view, the correct attainment involves removing the damaged cartilage, exposing the subchondral bone, penetrating the subchondral vascular area with clot formation at the defect level. The principle underlying these procedures is the transformation of fibrin from the clot into fibrocartilage, a cartilage-like tissue, but inferior in terms of mechanical and biological properties and wear resistance. The advantage is given by relative ease of accomplishment and reduced invasiveness (Figure 9.1.b);

- *Osteochondral transplantation* - also known as mozaicplasty, involves harvesting of cylinders formed from bone with healthy covering cartilage, from peripheral, healthy, non-weight bearing areas of the femoral trochlea, and their press-fit implantation in the receiving areas of the injured region. It is performed with special, single-use instruments of different sizes. In this technique, the screw-reattachment procedure of the detached osseocartilaginous fragment in case of osteochondritis dissecans or osteochondral fracture should be included, as well as the combination of the two methods: osteocondral cylinders and screws (Fig. 9.2.);

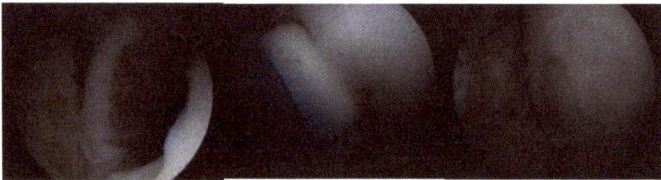

*Fig. 9.2. Osseocartilaginous transfer OATS*

- *Collagen membrane application over subchondral microfractures* - represents an extension of the microfracture technique by applying collagen membranes to the injured area such as the Chondro-Gide® reference membrane (*Geistlich Pharma AG, Switzerland*), a bilayer network of collagen I and III, which are sealed in position with a fibrin adhesive (*Tisseel Lyo®, Baxter, Deerfield, IL, USA*) (Figure 9.3);

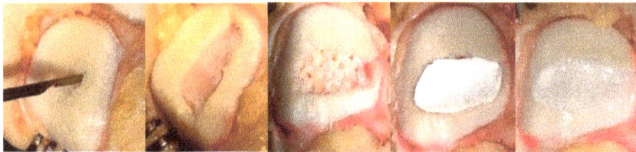

*Fig. 9.3. Chondral reconstruction with microfractures and collagen membrane*

- Hyaluronic acid membrane (Hyalofast®) application (Anika Therapeutics S.r.l., Padova, Italy) - following a similar procedure;

- *Intra-articular injection of multipotent stem cells from bone marrow concentrate (BMC)* - harvesting is performed by aspiration puncture from the iliac crest and centrifugation to obtain a bone marrow concentrate that is injected intraarticular at the desired area. The contained multipotent stem cells allow differentiation into osteocytes and chondrocytes, both embedded in the cartilage and the subchondral bone. This method is mainly used in multiple or diffuse chondral lesions (osteoarthritis).

The evaluation of the results was based on the analysis of the functional scores (*International Knee Documentation Committee (IKDC)*[169] subjective and objective evaluation of the knee and the *American Orthopaedic Foot and Ankle Society (AOFAS)*[170] for ankle assessment) performed pre- and post-operatively at 6 months for patients in the treatment groups. The statistical interpretation was based on the Student test for independent two-tailed variables. The statistical significance limit was established as $p < 0.05$.

## 9.4 Results

The population analysis of the distribution of cases according to the age of patients is shown in Table 9.2. The gender distribution is shown in Table 9.3. Next, in Table 9.4. the location of the lesions by group is shown. Group distributions of cases by grade (Outerbridge) and lesion area are shown in Tables 9.7 and 9.8. Related lesions and procedures are presented in Tables 9.9 and 9.10.

***Table 9.2.*** *Cases distribution in groups according to age.*

| Age group | 16-30 | 31-40 | 41-50 | 51-60 | 61-70 |
|---|---|---|---|---|---|
| **Group 1** | 5 | 8 | 10 | 9 | 5 |
| **Group 2** | 4 | 0 | 8 | 7 | 5 |
| **Group 3** | 4 | 1 | 1 | 1 | 0 |
| **Group 4** | 2 | 2 | 3 | 0 | 0 |
| **Total** | **15** | **11** | **22** | **17** | **10** |

***Table 9.3.*** *Cases distribution according to gender.*

| Gender | M | F |
|---|---|---|
| **Group 1** | 20 | 17 |
| **Group 2** | 16 | 8 |
| **Group 3** | 5 | 2 |
| **Group 4** | 4 | 3 |
| **Total** | **45** | **30** |

*Table 9.4. Cases distribution in groups according to localization (a single case could present more lesions with different localizations).*

| Localization | MFC | LFC | MTP | LTP | ROT | TROCH | TOTAL/GROUP |
|---|---|---|---|---|---|---|---|
| Group 1 | 23 | 3 | 5 | 3 | 22 | 8 | 64 |
| Group 2 | 20 | 3 | 2 | 1 | 8 | 4 | 38 |
| Group 3 | 4 | 2 | 0 | 0 | 2 | 1 | 9 |
| Group 4 | 2 | 4 | 0 | 1 | 2 | 0 | 9 |
| Total | 49 | 12 | 7 | 5 | 34 | 13 | 120 |

*Table 9.5. Cases distribution in groups according to lesional Outerbridge classification.*

| Stage | III | IV | TOTAL/GROUP |
|---|---|---|---|
| Group 1 | 21 | 16 | 37 |
| Group 2 | 13 | 11 | 24 |
| Group 3 | 1 | 6 | 7 |
| Group 4 | 1 | 6 | 7 |
| Total | 36 | 39 | 75 |

*Table 9.6. Cases distribution in groups according to defect area.*

| Area | < 1cm$^2$ | 1-2 cm$^2$ | 2-4 cm$^2$ | TOTAL/GROUP |
|---|---|---|---|---|
| Group 1 | 9 | 21 | 7 | 37 |
| Group 2 | 3 | 10 | 11 | 24 |
| Group 3 | 3 | 4 | 0 | 7 |
| Group 4 | - | 1 | 6 | 7 |
| Total | 15 | 36 | 24 | 75 |

*Table 9.7. Lesions associated to cartilage defects.*

| Associated lesions | No. of cases |
|---|---|
| Internal meniscus | 47 |
| External meniscus | 13 |
| ACL | 16 |
| Loose bodies | 9 |
| Hypertrophic synovitis | 12 |
| Genu valgum | 2 |
| Knee osteoarthritis | 14 |

*Table 9.8. Concomitant procedures.*

| Associated procedures | No. of cases | Percentage % |
|---|---|---|
| Meniscectomy | 55 | 92 |
| Meniscal suture | 8 | 13 |
| ACL ligamentoplasty | 16 | 100 |
| Loose bodies excision | 9 | 100 |
| Synovectomy | 12 | 100 |
| Correction osteotomy | 2 | 100 |

The analysis of the results of the subjective and objective IKDC knee scores revealed the aspects described below. For the numerical quantification of the objective IKDC score, which ranges from A to D, we noted group A with 4, B with 3, C with 2 and D with 1.

By comparing the subjective and objective IKDC scores between group of patients treated by lesion debridement or microfractures and patients treated with more advanced osteochondral reconstruction (OATS or collagen membrane application), statistically, significantly higher values are obtained for the last one in terms of IKDC objective postoperative scores (Table 9.9).

*Table 9.9. Comparison of IKDC scores between the groups with debridement or microfractures and those with OATS or collagen membrane.*

| Score | (Debridement + MFX) vs. (OATS + COLLAGEN) |
|---|---|
| IKDC S preop. | 0.01* |
| IKDC S postop. | 0.16 |
| IKDC O preop. | 0.053 |
| IKDC O postop. | 0.002* |

*statistically significant ($p < 0.05$)

Improvement of IKDC scores in patient groups treated with advanced chondral reconstruction techniques is highly significant statistically compared to the groups of patients treated with "classical" techniques (Table 9.10).

*Table 9.10. Comparison of pre and postoperative IKDC scores differences between the groups of patients.*

| Score | Mean differences: IKDC S postop. - IKDC S preop. | Mean differences: IKDC O postop. - IKDC O preop. |
|---|---|---|
| **Debridement** | 9.24** | 0.32** |
| **Microfractures** | 15.58** | 0.75** |
| **OATS** | 25.29** | 1.14** |
| **COLLAGEN** | 34.29** | 1.71** |

**highly statistically significant (p<0,001)

Absolute values of postoperative IKDC scores in all groups are higher than preoperative, the highest increase being recorded in group 4, of patients treated with collagen membranes (both subjective and objective scores) (Table 9.11). Starting from lower preoperative scores, groups 3 and 4 managed to increase them to the highest levels compared to groups 1 and 2 (Figure 9.4).

*Table 9.11. Comparison of mean IKDC scores values in pre and postoperative, between the groups of patients, altogether with standard deviations and confidence intervals.*

| Type of treatment | Mean values [standard deviation] / Interval | | | |
|---|---|---|---|---|
| | **IKDC S preop.** | **IKDC S postop.** | **IKDC O preop.** | **IKDC O postop.** |
| **1. Debridement** | 61.32 [7.95]/ [48-85] | 70.57 [9.13]/ [51-95] | 2.57 [0.55]/ [2-4] | 2.89 [0.57]/ [2-4] |
| **2. Microfractures** | 56.17 [12.97]/ [32-79] | 71.75 [13.37]/ [47-92] | 2.25 [0.53]/ [1-3] | 3 [0.66]/ [2-4] |
| **3. OATS** | 51.14 [16.68]/ [33-79] | 76.43 [13.45]/ [59-91] | 2.29 [0.49]/ [2-3] | 3.43 [0.53]/ [3-4] |
| **4. COLLAGEN** | 41.86 [15.37]/ [21-65] | 76.14 [12.05]/ [52-89] | 1.86 [0.69]/ [1-3] | 3.57 [0.53]/ [3-4] |

All patients in groups 3 and 4 went to a higher level of IKDC objective score. The highest increase per patient was achieved in group 4, with a value of 1.71 of the class/subject leap (increase of the objective IKDC score, considering A = 4 and D = 1), followed by group 3, with an average increase of 1.14 class/subject leap.

The lowest class leap was recorded in group 1 of debridement (mean 0.32/patient), followed by group 2 of microfractures (0.75/patient). In group 4, most jumps of 2 classes (71.4% of the group's patients) were recorded.

Postoperatively in groups 3 and 4 there are no more objective IKDC scores recorded from group C or D.

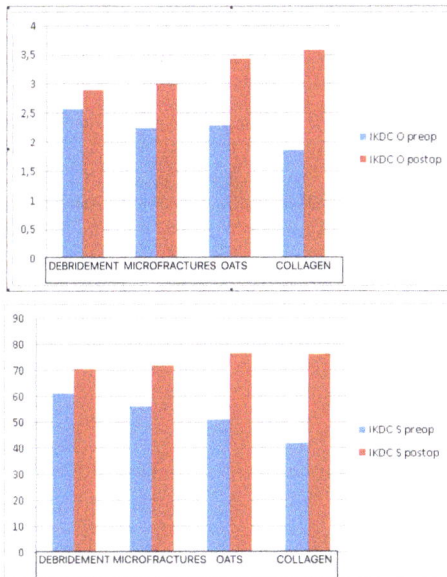

***Fig. 9.4.*** *Comparison between the evolution of scores IKDC S and O pre and postoperative among groups.*

Statistically significant correlations between the degree of chondral lesions (Outerbridge III or IV) and the IKDC scores could not be established. On the other hand, there are slightly negative correlation between the patient's age and the functional (pre-and postoperative) scores, and between the size of the lesions and the functional scores (Table 9.12), moderate negative correlation between the subjective and objective IKDC postoperative scores and patient's age.

*Table 9.12. Correlations of IKDC pre/postoperative scores.*

| Score | Correlation: lesion size – functional score | Correlation: age – functional score |
|---|---|---|
| IKDC S preop. | -0.29* | -0.33* |
| IKDC S postop. | -0.24 | -0.54** |
| IKDC O preop. | -0.34* | -0.35* |
| IKDC O postop. | -0.19 | -0.55** |

\* Negative slight correlation; \*\* Negative moderate correlation.

There were no serious complications in the follow-up of patients, as there were only 27 cases of post-procedural local hematomas, 3 cases of deep vein thrombosis, which had a very good evolution under treatment and did not significantly influence the healing processes and medical recovery.

### 9.4.1    Particular clinical cases

*CASE 1. K.T., ♂, 22 y.o.*

**Diagnosis:** Stage IV post-traumatic chondropathy of the right lateral femoral condyle (Figure 9.5). Posterior medial meniscus tear of the knee following ACL ligamentoplasty (performed 4 years ago).

**Surgical intervention**: Chondroplasty of the right lateral femoral condyle with collagen I/III chondral implant (*Chondro-Gide*®), associated with subchondral microfractures (AMIC technique - Autologous Matrix Induced Chondrogenesis). Arthroscopic all-inside repair of the posterior medial meniscus of the right knee with a Fast-Fix® implant.

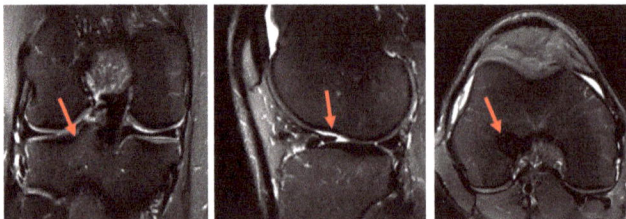

*Fig. 9.5. Stage IV chondropathy of lateral femoral condyle.*

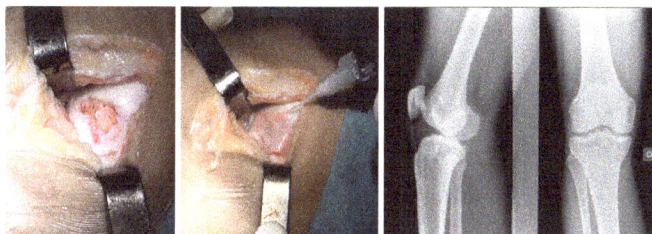

***Fig. 9.6.*** *Intraoperative aspect and postoperative radiographies.*

***Results:*** Starting from an extensive chondral lesion of approximately 4 cm$^2$, IKDC subjective score improved from 65 preoperatively, to 89 postoperatively at 6 months. Also, group B was transferred to group A at objective IKDC score, obtaining a mobile and painless knee that allowed to resume to physical activities without noticeable complications (Figure 9.6).

*CASE 2. S.F., ♂, 23 y.o.*

***Diagnosis:*** Chronic instability of the left knee due to ACL lesion. Stage IV post-traumatic chondropathy of the left lateral femoral condyle (Figure 9.7.a). Complex tears of the medial and lateral meniscus of the left knee following ACL ligamentoplasty (performed 4 years ago).

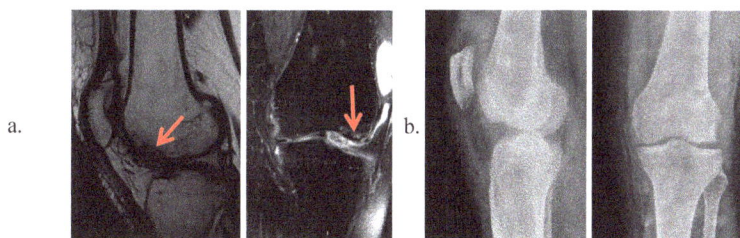

***Fig. 9.7.*** *a. Stage IV chondropathy of lateral femoral condyle; b. Cartilage reconstruction.*

***Surgical intervention:*** Left ACL arthroscopic reconstruction with semitendinosus/ gracilis (STG) tendon autograft. Chondroplasty of the left lateral femoral condyle with collagen I/III chondral implant (*Chondro-Gide*®), associated with subchondral microfractures (AMIC technique). Arthroscopic meniscectomy of the medial and lateral menisci of the left knee (Figure 9.7.b).

**Results:** Starting from an extensive chondral lesion of approximately 4 cm$^2$, due to an unstable knee and associated meniscal tear, the IKDC subjective score improved from 44 preoperatively to 81 postoperatively at 6 months. Also, group C was transferred to group A at objective IKDC score, achieving a stable, mobile and painless knee that allowed the return to physical activities without significant complications (only a persistent effusion for 6 weeks postoperatively, submitted under physio-kinesiotherapy).

*CASE 3. L.I., ♀, y.o.*

**Diagnosis**: Stage IV degenerative right patellar chondropathy (Figure 9.8). Stage IV chondropathy of the right lateral femoral condyle on the bearing surface. Right patellofemoral dysplasia.

**Surgical intervention:** Patellar chondroplasty of the left knee with hyaluronic acid implant (*Hyalofast*$^®$), associated with subchondral microfractures (AMIC technique). Right lateral femoral condyle mozaicplasty (OATS$^®$ technique - Osteochondral Autograft Transfer System, 8 mm diameter). Lateral patellar retinaculum release.

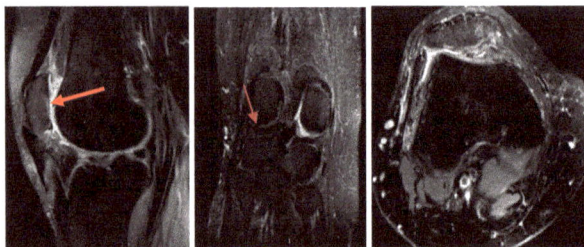

**Fig. 9.8.** *Stage IV chondropathy of the patella and lateral femoral condyle.*

**Results:** Starting from an extensive chondral lesion of approximately 4 cm$^2$, IKDC subjective score improved from 21 preoperatively to 52 postoperatively at 6 months. It was transferred from group D to group B in objective IKDC score, knee mobility being resumed, but with the persistence of pain and cracking at flexion/extension movements, especially when climbing/descending the stairs, the patient also accusing the sensation of articular blockage.

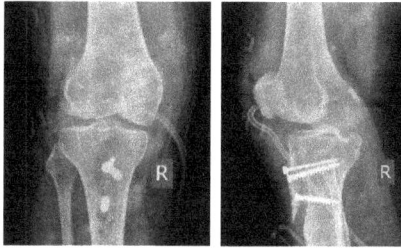

***Fig. 9.9.*** *Stage IV chondropathy on the patella and lateral femoral condyle, operated by TTA transposition.*

For these reasons, 12 months after the first operation, another surgical intervention was performed by translating the anterior tibial tuberosity by its antero-medialization (the Fulkerson procedure), lateral patellar release and medial patellar retinaculum plication and restoring the patellar tracking (modified due to external patellar subluxation due to patellofemoral dysplasia) (Figure 9.9). After this intervention, the symptoms of the patient diminished significantly, returning to a painless status after the balneo-physio-kinesiotherapy rehabilitation treatment.

*CASE 4. B.M., ♀, 44 y.o.*

***Diagnosis***: Vicious consolidation of a lateral tibial plateau depression fracture (operated in another orthopedic service), on the left knee, with genu valgum deviation and deterioration of the external tibial articular surface and secondary knee instability (Figure 9.10).

***Surgical intervention:*** Reconstruction of the left external tibial plateau joint with cortico-cancellous bone graft harvest from the iliac crest and applying a collagen I/III chondral implant (*Chondro-Gide*[R]). Lateral open-wedge high tibial osteotomy and osteosynthesis with an L-shaped LCP plate with 7 screws and cortico-cancellous bone graft application (Figures 9.11.a and b). Osteotomy of the left fibular diaphysis in the middle third.

***Results:*** Starting from an extensive chondral lesion of approximately 4 cm$^2$, with a painful, unstable, stiff and valgus malaligned knee, IKDC subjective score improved from 23 preoperatively to 78 postoperatively at 6 months. Also, group D was transferred to group B in objective IKDC score, obtaining a corrected mechanical axis, stable, mobile and painless knee, which allowed the return to physical activities without noticeable complications.

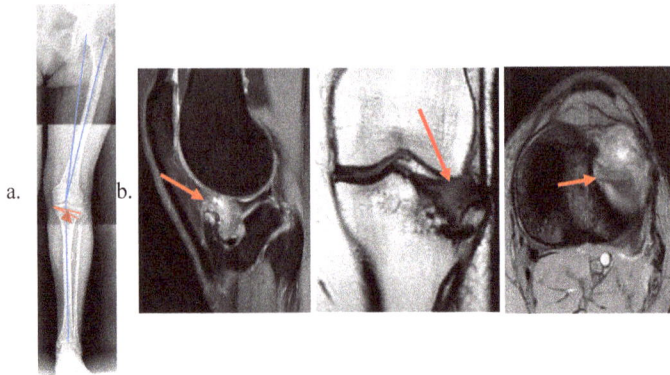

***Fig. 9.10.*** *Osseocartilaginous defect of the lateral tibial plateau with genu valgum malalignment.*

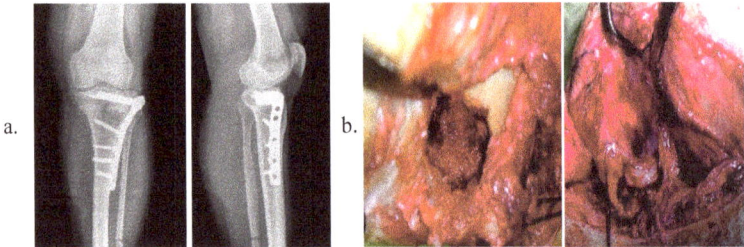

***Fig. 9.11.a.*** *Radiologic aspect of the reconstruction at 6 months postoperative;* ***b.*** *Intraoperative aspect.*

*CASE 5. N.B., ♀, 35 y.o.*

***Diagnosis***: Stage III degenerative right patellar chondropathy (Figure 9.12 a). Anterior horn of the medial meniscus tear in the right knee.

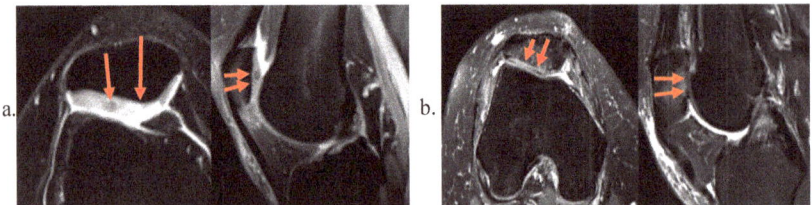

***Fig. 9.12.a.*** *Preoperative MRI aspect;* ***b.*** *Postoperative MRI aspect at 4 months.*

**Surgical intervention:** Patellar chondroplasty of the right knee with hyaluronic acid implant (*Hyalofast*®), associated with subchondral microfractures (AMIC technique) (Figure 9.3). Arthroscopic regularization of the anterior horn of the right medial meniscus.

**Results:** Starting from an extensive chondral lesion of approximately 4 $cm^2$, associated with a patellofemoral pain syndrome, IKDC subjective score improved from 45 preoperatively to 70 postoperatively at 6 months. Also, group C was transferred to group B at objective IKDC score, obtaining a mobile and painless knee, which allowed the return to physical activities after physiotherapy without significant complications (Figure 9.12 b).

*CASE 6. B.R., ♂, 35 y.o.*

**Diagnosis**: Diagnosis: Chronic instability of the right knee due to ACL lesion. Stage IV post-traumatic chondropathy of the right internal femoral condyle (Figure 9.13)

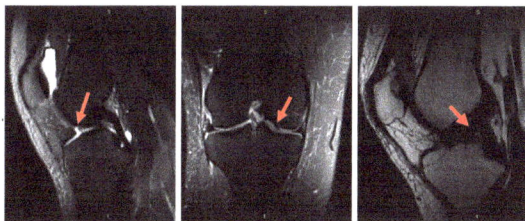

**Fig. 9.13.** *Stage IV chondropathy of the medial femoral condyle, on ancient ACL lesion.*

**Surgical    intervention:**    Right    ACL    arthroscopic    reconstruction    with semitendinosus/gracilis (STG) tendon autograft. Chondroplasty of the right internal femoral condyle with collagen I/III chondral implant (*Chondro-Gide*®), associated with subchondral microfractures (AMIC technique) (Figure 9.14).

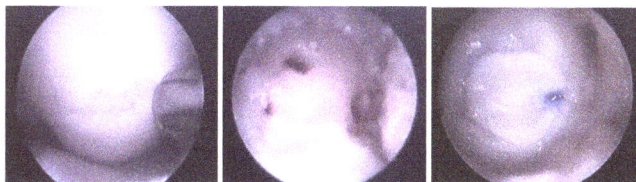

**Fig. 9.14.** *Debridement, microfractures and collagen membrane application.*

***Results:*** Starting from a chondral lesion of approximately 2 cm$^2$, on an unstable knee, the IKDC subjective score improved from 45 preoperatively to 82 postoperatively at 6 months. It was also transferred from Group C to Group A in objective IKDC score, achieving a stable, mobile and painless knee that allowed the return to physical activities after physiotherapy, without significant complications (Figure 9.15).

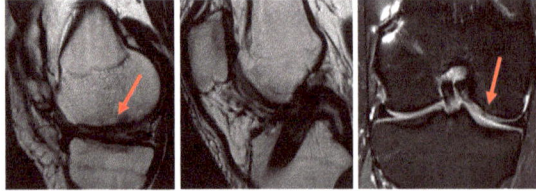

**Fig. 9.15.** *Control MRI at 6 months follow-up with almost complete healing of cartilage.*

*CASE 7. C.T., ♂, 43 y.o.*

***Diagnosis***: Stage IV degenerative chondropathy of the right medial femoral condyle. Posterior horn medial meniscus tear at the right knee.

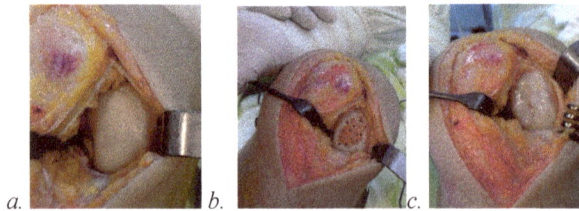

**Fig. 9.16.** *Reparation of the chondral lesion with collagen membrane and subcondral microfractures.*

***Surgical intervention:*** Chondroplasty of the right medial femoral condyle with collagen I/III chondral implant (*Chondro-Gide*®), associated with subchondral microfractures (AMIC technique) (Figure 9.16). Arthroscopic regularization of the posterior horn of the right medial meniscus.

***Results:*** Starting from an extensive chondral lesion of approximately 4 cm$^2$, the IKDC subjective score improved from 50 preoperatively to 81 postoperatively at 6 months. Also, group C was transferred to group A in objective IKDC score, resulting in a painless knee, which allowed the return to physical activities with a slight limitation of flexion to 120 degrees.

*CASE 8. B.G., ♂, 37 y.o.*

**Diagnosis**: Stage IV post-traumatic chondropathy of the left tibio-talar joint, with secondary arthrosis (Figure 9.17).

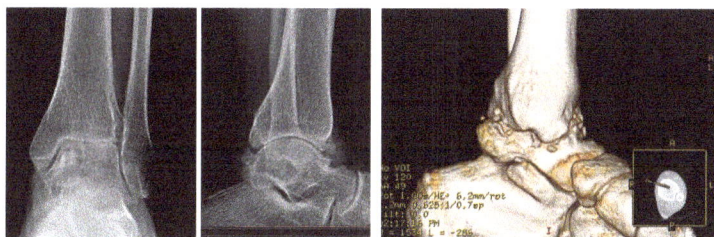

**Fig. 9.17.** *Generalized chondropathy of the ankle with arthritic changes.*

**Surgical intervention:** Intra-articular ankle injection of bone marrow aspirate concentrate from the iliac crest (BMAC® - Harvest® Bone Marrow Aspirate Concentrate, Harvest Terumo BCT Europe N.V, Zaventem, Belgium).

**Results**: Starting from an almost complete damaged cartilage of the left tibiotalar joint, following multiple post-traumatic osteochondral lesions, the AOFAS score was improved from 48 preoperatively to 60 postoperatively at 12 months (from a possible maximum of 100). The control MRI images at 12 months post-injection showed a slight expansion of the tibiotalar joint space in the anterior third (height above 2 mm, compared with 1.6 mm previously) and a better visualization of the remaining cartilage at this level, as it can be seen in Figure 9.18. Due to the generalized chondral destruction with advanced arthrosis, significant improvement in the AOFAS functional score was not achieved.

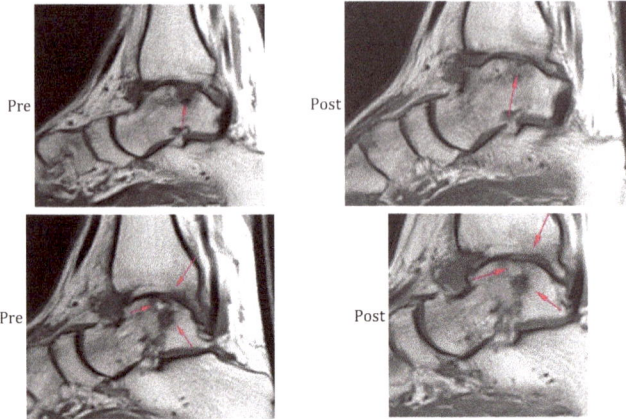

***Fig. 9.18.*** *Increase of articular cartilage height at 12 months post-intervention.*

However, this case exemplifies the regenerative potential of multipotent stem cell concentrates, which are a promise for the recovery of articular cartilage, of course, up to certain limits, while endoprosthetic replacement remains the only successful therapeutic alternative in extensive chondral destruction.

*CASE 9. L.G., ♂, 53 y.o.*

***Diagnosis***: Stage IV osteochondritis dissecans of the left medial femoral condyle (Fig. 9.19).

***Surgical intervention:*** Combined chondroplasty by lesional debridement, subchondral drilling, and reinsertion of the chondral fragment with 2 cannulate screws plus an osteochondral cylinder transfer (Figures 9.20 and 9.21).

***Results:*** Starting from a chondral lesion of approximately 2 cm$^2$ located at the bearing surface of the medial femoral condyle, the IKDC subjective score improved at 6 months from 43 preoperatively to 70 postoperatively. It was also transferred from group C to group B at objective IKDC score. After the recovery treatment, it was possible to restore a painless, mobile joint with excellent functionality.

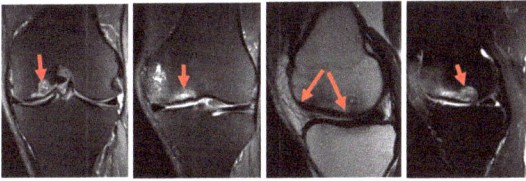

*Fig. 9.19. Osteochondritis dissecans stage IV on the medial femoral condyle*

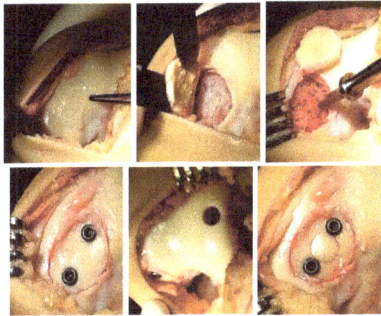

*Fig. 9.20. Debridement, chondral fragment refixation with 2 screws + 1 OATS cylinder.*

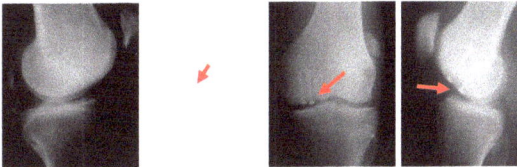

*Fig. 9.21. Pre and postoperative radiologic control.*

*CASE 10. C.S., ♂, 26 y.o.*

***Diagnosis***: Stage III osteochondritis dissecans of the talus (Figure 9.22).

***Surgical intervention:*** Arthroscopic chondroplasty of the osteochondral lesion of the talus through debridement and subchondral microfractures.

***Results:*** Starting from a chondral lesion of approximately 1 cm$^2$, located at the tibiotalar joint surface, the AOFAS score improved at 6 months from 70 preoperatively to 80 postoperatively. After the recovery treatment, it was possible to restore a painless, mobile joint with excellent functionality.

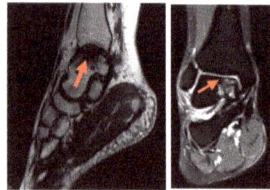

*Fig. 9.22. Stage III talar osteochondritis dissecans*

## 9.5     Discussions

Cartilage lesions are very common in clinical practice, but only a small amount of them requires specialized treatment. The incidence reported by us corresponds to the data reported in the literature[46-49,129].

The main limitation of our study is that it is a retrospective, observational study. But even so some aspects are relevant. It is clearly demonstrated the superiority of advanced treatments (mozaicplasty, collagen membranes) in relation to classical treatments (debridement and microfractures), although the last ones are widely practiced as technically they are much easier. Even if the comparisons are not strictly unitary, respectively on the exact same type of lesions, the statistical significance is high.

Table 9.10 shows a significant improvement in IKDC scores for patients in groups 3 and 4. Of course, we are aware of the fact that these groups have been treated having progressed lesions, thus starting at much lower values. In relation to the number of subjects in the groups, the highest proportion of those passing to a higher level at IKDC score is met in groups 3 and 4, in which there are otherwise no patients with a group C or D objective score. Also in these groups are registered the highest frequency of leaps over two classes of assessment of the objective IKDC score (Figure 9.4).

Patients who underwent debrided have the lowest increase in functional knee score, both objective and subjective. Particular cases best exemplify the immense possibilities that biomaterial engineering has. The modern collagen implants used allow the reconstruction of very large chondral defects, practically achieving a kind of biological arthroplasty that has the role of preventing arthritic degradation.

## 9.6     Conclusions

> Chondral lesions are encountered with a very high frequency, but only a small amount requires specialized treatment (around 17%).

> Cartilage lesions should be considered in any joint disease requiring treatment (meniscal lesions, ligamentous lesions, etc.).

➢ Debridement and subchondral microfractures do not bring the expected benefit in most cases.

➢ The more complex procedures bring the greatest benefits and offer the possibility of restoring large areas of articular cartilage (about 4 $cm^2$).

➢ Osteochondral implants provide far better results than implant-free procedures, especially in combination with multipotent stem cells.

➢ Tissue engineering is the benchmark for cartilage repair procedures.

## 10.   General Conclusions

Following our study on the isolation of bone marrow stem cells (MSCs) from the iliac crest (BMC on an ovine animal model) using the Concemo® kit and the evaluation of cell function on demineralized MatriBone Ortho biomaterial, we can formulate the following conclusions:

• Hematogenic bone marrow and adipose tissue isolated cells can be easily processed using the Concemo® kit, but the immunophenotypic results indicate a fairly wide variability between the harvested and analyzed samples;

• Until the first passage, cell cultures showed a marked heterogeneity, with the predominance of round and fusiform cells, with the confluence of cell cultures on average after 14 days from isolation;

• After the first passage, positivity was recorded for the CD44 mesenchymal marker, the global analysis indicating an average of 96.67%;

• The degree of attachment and proliferation on MatriBone Ortho demineralized was significantly higher compared to control cell culture;

• The cells show a comparable degree of attachment on the B3 implant to the B4 implant (Chondro-Gide), which is currently a reference implant for the treatment of articular cartilage defects;

• The results of the migration potential assessment on demineralized B3 show statistically significantly better results compared to the control culture;

• We recommend the cultivation and transplantation of mesenchymal stem cells on the B3 (MatriBone Ortho) implant, which is biocompatible, does not exhibit residual cytotoxicity and allows good adherence and proliferation of mesenchymal stem cells.

The study concerning the new way of separating ASCs stem cells from lipoaspirate fluid allows us to formulate the following conclusions:

• ASCs is a valuable option for osteochondral repair, providing at least the same good results as the processed lipoaspirate (PLA) cells for chondral repair;

• LAF-derived stem cells appear to have somewhat better activity and effects compared to the PLA cells, as regards the repair of the cartilage, being aided by the trophic molecules in the liquid;

• LAF cells can be quickly separated by minimal tissue manipulation, making them cheaper and more suitable for single-step surgical procedures;

Multidisciplinary research approach is the premise of obtaining the best results for the development of complex therapies in osteoarticular disorders.

From the comparative complex study of the efficacy of treatment with a new type of collagen implant and bone marrow stem cells versus lipoaspirate cells, the following results were obtained:

• The sheep is a very good animal model for therapeutic experiments on the locomotor apparatus, in particular on the treatment of osteochondral lesions, which can go further for studying degenerative joint diseases (osteoarthritis);

• Methods for bone marrow stem cell extraction from the iliac crest as well as adipose tissue are also feasible in sheep, resulting in multipotent mesenchymal cell populations with osteochondral regeneration potential;

• The tendency for centripetal bone neo-formation can be observed, starting from the walls of the defect, with the formation of a nearly complete subchondral bridge;

• Instead, it can be seen in some cases the deepening of the lesion towards the deep area of the defect, in the form of an extension of the defect created or a pseudocyst, which is also very well observed on the microscopic sections and then explained on the 3D virtual model through the local distribution of the forces;

• Regeneration of hyaline cartilaginous tissue is possible with tissue of the same quality, not just fibrocartilage;

• Procedures involving repair in one time, without cell manipulation and laboratory processing are the most effective in terms of the complexity of the therapeutic act but also from an economic point of view;

• Mesenchymal stem cells offer real possibilities for treatment of focal cartilage defects, with or without the involvement of the underlying bone;

• Mesenchymal stem cells from BMC and ASC clearly demonstrate their superiority to local mesenchymal cells in the subchondral layer;

• For better efficacy, it is necessary to use implants that provide support for these cells, but also to restore the architecture of collagen fibers from bone and cartilage;

• The collagen implant tested for the first time meets the characteristics required to be used as a support for osteochondral regeneration, providing very good histological, imagistic and clinical results;

• BMC cells provide better results in the regeneration of bone and cartilage than ASC cells, but these good results should be also considered due to their relative ease in harvesting and processing them. Their activation with autologous concentrated platelet-rich plasma (PRP) increases their repair potential;

• The results of the study are encouraging, showing the immense possibilities of research that this field offers, but it is necessary to continue with larger experimental studies on larger animal models of therapeutic methods and to assess their wider applicability on clinical scale;

• The use of high-performance imaging techniques to evaluate results, such as MRI, CT, should be encouraged, thereby avoiding the sacrifice of animals;

• Immunohistochemical investigations allow the assessment of quality of the repair tissue by detecting the type II collagen belonging to cartilaginous tissue;

• It is necessary to find simple, reproducible methods with minimal morbidity and risks, cheaper, but with better results and reduced complications.


The mathematical modeling of the sheep's knee joint based on the CT reconstruction allows us to test several scenarios with different loading forces, drawing some conclusions:

• In the simulated scenarios with 1120 N (normal) and 2240 N (excessive) forces, it is demonstrated that there are no stresses the cortical bone of the femur or tibia that is close to its fracture limit, in reality being below 50 MPa;

• From FEA simulations on Case 0 - healthy, it turns out that cartilage behaves like a compressible and elastic material that absorbs stresses and distributes them into adjacent bone tissue. This cartilage behavior is similar to that of the periodontal ligament (part of the dental tissues). After applying an excessive static force, the healthy femoral cartilage is loaded with a stress of 3 MPa at the contact with the tibial cartilage. Similar results have been obtained in Case 2, where the cartilage has a stress ranging between 0.5-3 MPa, a stress that is considered normal for cartilage contact;

• After applying the treatment in Case 2, the cartilage is restored 6 months postoperatively, and normal movements distribute stresses into tissue similar to Case 0 - healthy. Complex movements involving high forces could affect regions in the restored trabecular bone;

• The existing collagen in the defect at 6 months is stress loaded from the cortical bone and it distributes directly into the other tissues with which it is in contact, reaching to transmit high stresses in both the trabecular bone of the femur and in cartilage. This phenomenon could particularly damage the lateral walls of the newly formed cartilage of the femur, but also the tibial cartilage with which they come into direct contact, visible in Cases 1 and 3 (Figure 6.11 and 6.13);

• The behavior of collagen to lead stresses is explained by its rigidity compared to elastic tissues (spongy bone and cartilage). In Cases 1 and 3, after applying a force of 2240 N, the newly formed cartilage in the defect area is stress loaded in some areas reaching at 5-7 MPa. This level of stress can damage the cartilage. The evidences available in the literature suggests that some disorders with cartilage damage are associated with cartilage stress between 1-20 MPa;

• The existence of collagen in the defect area leads to an increase in deformations both normal situations (1-2 mm) and after applying an excessive pressure (1-3 mm). When assessing the biomechanical response of cartilage of the animals treated in vivo, FEA studies may help to make clinical decisions about how and what kind of materials can be applied in the treatment of human cases, knowing that without effective intervention, the progressive loss of the affected cartilage may be promoted.

From the retrospective clinical study, the following results were obtained:

• Chondral lesions are encountered with a very high frequency, but only a small amount requires specialized treatment (around 17%);

• Cartilage lesions should be considered in any joint disease requiring treatment (meniscal lesions, ligamentous lesions, etc.);

• Debridement and subchondral microfractures do not bring the expected benefit in most cases;

• The more complex procedures bring the greatest benefits and offer the possibility of restoring large areas of articular cartilage (about 4 $cm^2$);

• Osteochondral implants provide far better results than implant-free procedures, especially in combination with multipotent stem cells;

• Tissue engineering is the benchmark for cartilage repair procedures.

Although the results obtained so far with stem cells are very good and encouraging, it is necessary to find a method of their use for the treatment of extensive lesions over 4 $cm^2$, and prevention of arthritic degradation and its generalization.

These studies give us an insight into the future in the hope of finding an effective remedy for degenerative diseases.

## Acknowledgements

The development of the research within the doctoral studies that led to the elaboration of this book were not possible without permanent surveillance and coordination of Prof. Gheorghe TOMOAIA, MD, PhD, from the Orthopedics Discipline of UMF "Iuliu Hatieganu" Cluj-Napoca, Romania, to whom I express my acknowledgement and gratitude for his trust, advices and encourragements, as teacher and menthor.

I am also thankful to Prof. Habil. Iulian ANTONIAC, PhD, from the Faculty of Materials Science and Engineering, University Polytehnica of Bucharest, Romania, for his shared expertise in the field of biomaterials enginnering, support for ellerationg the manuscript and great networking, who encouraged me to reach the envisaged goals.

The secret of success is represented by a team multidisciplinary approach, reason for what I wish to thank from my heart to all those who helped me in accomplishing the research, and in particular to Prof. Aurel MUSTE, MD, PhD and his team from Surgery Clinic of University of Agricultural Sciences and Veterinary Medicine USAMV of Cluj-Napoca, Emoke PALL, MD, PhD and Mihai CENARIU, MD, PhD from Reproduction Discipline of USAMV Cluj-Napoca, Dan GHEBAN, MD, PhD and Rodica BOROS from the Pathological Anatomy Discipline of UMF Cluj-Napoca, Prof. Septimiu TOADER, MD, PhD and Cristian BERCE, MD, PhD from UMF "Iuliu Hatieganu" Biobase, Prof. Maria TOMOAIA-COTISEL, PhD, Roxana PASCA, PhD and the whole team of Atomic Force Microscopy Laboratory from Physical-Chemistry Center of Babes-Bolyai University UBB Cluj-Napoca, Olga Soritău, MD, PhD from "Ion Chiricuta" Oncology Institute IOCN Cluj-Napoca, Assoc. Prof. Lucian BARBU-TUDORAN, PhD and his team from Electron-Microscopy Center of Biology Faculty of UBB Cluj-Napoca, Prof. Simion SIMON, PhD and Alexandru FARCASANU from National Center of Magnetic Resonance (CNRM) of Physics Faculty, UBB Cluj-Napoca, and not least to Eng. Cosmin COSMA, PhD, from the Department of Fabrication Engineering of Technical University UTCN of Cluj-Napoca, who guided me for elaborating the statical study with finite elements of osteo-articular modelling.

Not least I would like to express my gratitude to my team of collaborators from the University Clinic of Orthopedics and Traumatology for their help in the clinical activity and research, naming here Dragos APOSTU, MD, PhD, Zsolt GABRI, MD and Daniel OLTEAN-DAN, MD, PhD.

I also thank my beloved family for their understanding and support, and I must say it's only them who know what that means… .

And for all those maybe not named here… .

## References

[1]     Camarero-Espinosa S, Rothen-Rutishauser B, Foster EJ, Weder C, Vila G, Caceres E, et al. Articular cartilage: from formation to tissue engineering. Biomater Sci . 2016 [cited 2017 Mar 20];4(5):734–67. Available from: http://xlink.rsc.org/?DOI=C6BM00068A

[2]     Bayliss MT, Osborne D, Woodhouse S, Davidson C. Sulfation of chondroitin sulfate in human articular cartilage. The effect of age, topographical position, and zone of cartilage on tissue composition. J Biol Chem.1999 May 28 [cited 2017 Mar 20];274(22):15892–900. Available from: http://www.ncbi.nlm.nih.gov/ pubmed/10336494

[3]     Correa D, Lietman SA. Articular cartilage repair: Current needs, methods and research directions. Semin Cell Dev Biol [Internet]. 2016; Available from: http://dx.doi.org/ 10.1016/j.semcdb.2016.07.013

[4]     Wang Y, Wei L, Zeng L, He D, Wei X. Nutrition and degeneration of articular cartilage. Knee Surgery, Sport Traumatol Arthrosc. 2013 Aug 4 [cited 2017 Mar 7];21(8):1751–62. Available from:http:// link.springer.com/10.1007/s00167-012-1977-7

[5]     Wong M, Carter DR. Articular cartilage functional histomorphology and mechanobiology: a research perspective. Bone [Internet]. 2003 Jul [cited 2017 Mar 7];33(1):1–13. Available from: http://www.ncbi.nlm.nih.gov/pubmed/12919695

[6]     Pazzaglia UE, Beluffi G, Benetti A, Bondioni MP, Zarattini G. A Review of the Actual Knowledge of the Processes Governing Growth and Development of Long Bones. Fetal Pediatr Pathol [Internet]. 2011 Mar 28 [cited 2017 Mar 7];30(3):199–208. Available from: http://www.ncbi.nlm.nih.gov/pubmed/21355682

[7]     Freemont AJ, Hoyland J. Lineage plasticity and cell biology of fibrocartilage and hyaline cartilage: Its significance in cartilage repair and replacement. Eur J Radiol [Internet]. 2006 Jan [cited 2017 Mar 7];57(1):32–6. Available from: http://linkinghub.elsevier.com/ retrieve/pii/S0720048X05002901

[8]     Carbonell-Blasco P, Antoniac I, Martin-Martinez MJ. New polyurethane sealants containing rosin for non-invasive disc regeneration surgery. Book Series: Key Engineering Materials. 2014;583:67-79.

[9]     Pelttari K, Pippenger B, Mumme M, Feliciano S, Scotti C, Mainil-Varlet P, et al. Adult human neural crest-derived cells for articular cartilage repair. Sci Transl Med [Internet]. 2014 Aug 27 [cited 2017 Mar 7];6(251):251ra119-251ra119. Available from: http://www.ncbi.nlm.nih.gov/pubmed/25163479

[10]    Yang C-Y, Chanalaris A, Troeberg L. ADAMTS and ADAM metalloproteinases in osteoarthritis – looking beyond the "usual suspects." Osteoarthr Cartil [Internet]. 2017 Feb 13 [cited 2017 Mar 8]; Available from: http://www.ncbi.nlm.nih.gov /pubmed/28216310

[11]    Kemp SF, Mutchnick M, Hintz RL. Hormonal control of protein synthesis in chick chondrocytes: a comparison of effects of insulin, somatomedin C and triiodothyronine.

Acta Endocrinol (Copenh) [Internet]. 1984 Oct [cited 2017 Mar 8];107(2):179–84. Available from: http://www.ncbi.nlm.nih.gov/pubmed/6093416

[12]	Raj A, Wang M, Liu C, Ali L, Karlsson NG, Claesson PM, et al. Molecular synergy in biolubrication: The role of cartilage oligomeric matrix protein (COMP) in surface-structuring of lubricin. J Colloid Interface Sci [Internet]. 2017 Jun 1 [cited 2017 Mar 8];495:200–6. Available from: http://www.ncbi.nlm.nih.gov/pubmed/28208081

[13]	Gao Y, Liu S, Huang J, Guo W, Chen J, Zhang L, et al. The ECM-Cell Interaction of Cartilage Extracellular Matrix on Chondrocytes. Biomed Res Int [Internet]. 2014 [cited 2017 Mar 8];2014:1–8. Available from: http://www.ncbi.nlm.nih.gov/pubmed/24959581

[14]	Eyre D. Collagen of articular cartilage. Arthritis Res [Internet]. 2002 [cited 2017 Mar 8];4(1):30. Available from: http://www.ncbi.nlm.nih.gov/pubmed/11879535

[15]	Newman AP. Articular Cartilage Repair. http://dx.doi.org/101177/0 3635465980260022701 [Internet]. 2016 [cited 2017 Mar 8]; Available from: http://journals.sagepub.com/doi/abs/ 10.1177/03635465980260022701

[16]	Bhosale AM, Richardson JB. Articular cartilage: structure, injuries and review of management. Br Med Bull [Internet]. 2008 Aug 1 [cited 2017 Mar 8];87(1):77–95. Available from: https://academic.oup.com/bmb/ article-lookup/doi/10.1093/bmb/ldn025

[17]	Deshmukh S, Dive A, Moharil R, Munde P. Enigmatic insight into collagen. J Oral Maxillofac Pathol [Internet]. 2016 [cited 2017 Mar 8];20(2):276. Available from: http://www.ncbi.nlm.nih.gov/pubmed/27601823

[18]	Deng H, Huang X, Yuan L. Molecular genetics of the COL2A1-related disorders. Mutat Res Mutat Res [Internet]. 2016 Apr [cited 2017 Mar 8];768:1–13. Available from: http://www.ncbi.nlm.nih.gov/pubmed/27234559

[19]	Barat-Houari M, Sarrabay G, Gatinois V, Fabre A, Dumont B, Genevieve D, et al. Mutation Update for COL2A1 Gene Variants Associated with Type II Collagenopathies. Hum Mutat [Internet]. 2016 Jan [cited 2017 Mar 8];37(1):7–15. Available from: http://www.ncbi.nlm.nih.gov/pubmed/26443184

[20]	Hardingham TE, Fosang AJ. Proteoglycans: many forms and many functions. FASEB J [Internet]. 1992 Feb 1 [cited 2017 Mar 8];6(3):861–70. Available from: http://www.ncbi.nlm.nih.gov/pubmed/1740236

[21]	Vynios DH, H. D. Metabolism of cartilage proteoglycans in health and disease. Biomed Res Int [Internet]. 2014 [cited 2017 Mar 8];2014:452315. Available from: http://www.ncbi.nlm.nih.gov/pubmed/25105124

[22]	Aspberg A. The Different Roles of Aggrecan Interaction Domains. J Histochem Cytochem [Internet]. 2012 Dec [cited 2017 Mar 8];60(12):987–96. Available from: http://journals.sagepub.com/doi /10.1369/0022155412464376

[23]	Rédini F. [Structure and regulation of articular cartilage proteoglycan expression]. Pathol Biol (Paris) [Internet]. 2001 May [cited 2017 Mar 8];49(4):364–75. Available from: http://www.ncbi.nlm.nih.gov/pubmed/11428173

[24]   Buraschi S, Neill T, Goyal A, Poluzzi C, Smythies J, Owens RT, et al. Decorin causes autophagy in endothelial cells via Peg3. Proc Natl Acad Sci U S A [Internet]. 2013 Jul 9 [cited 2017 Mar 8];110(28):E2582-91. Available from: http://www.ncbi.nlm.nih.gov/pubmed/23798385

[25]   Roughley P. The structure and function of cartilage proteoglycans. Eur Cells Mater [Internet]. 2006 Nov 30 [cited 2017 Mar 8];12:92–101. Available from: http://ecmjournal.org/journal/papers/vol012/ pdf/v012a11.pdf

[26]   Garvican ER, Vaughan-Thomas A, Clegg PD, Innes JF. Biomarkers of cartilage turnover. Part 2: Non-collagenous markers. Vet J [Internet]. 2010 Jul [cited 2017 Mar 8];185(1):43–9. Available from: http://linkinghub.elsevier.com/retrieve/pii/S1090023310001206

[27]   Lewis R, Feetham CH, Barrett-Jolley R. Cell Volume Regulation in Chondrocytes. Cell Physiol Biochem. 2011 [cited 2017 Mar 20];28:1111–22. Available from: www.karger.com

[28]   Vergroesen P-PA, Kingma I, Emanuel KS, Hoogendoorn RJW, Welting TJ, van Royen BJ, et al. Mechanics and biology in intervertebral disc degeneration: a vicious circle. Osteoarthr Cartil. 2015 Jul [cited 2017 Mar 20];23(7):1057–70. Available from: http://www.ncbi.nlm.nih.gov/ pubmed/25827971

[29]   Hunziker EB, Kapfinger E, Geiss J. The structural architecture of adult mammalian articular cartilage evolves by a synchronized process of tissue resorption and neoformation during postnatal development. Osteoarthr Cartil. 2007 Apr [cited 2017 Mar 21];15(4):403–13. Available from: http://www.ncbi.nlm.nih.gov/ pubmed/17098451

[30]   Klein MS TJ, Schumacher BS BL, Schmidt MS TA, Li KW, Voegtline MS, Masuda K, et al. Tissue engineering of stratified articular cartilage from chondrocyte subpopulations. [cited 2017 Mar 21]; Available from: http://www.oarsijournal.com/article/S1063-4584(03)00090-6/pdf

[31]   Amanatullah DF, Yamane S, Reddi AH. Distinct patterns of gene expression in the superficial, middle and deep zones of bovine articular cartilage. J Tissue Eng Regen Med . 2012 Jul [cited 2017 Mar 21];8(7):n/a-n/a. Available from: http://www.ncbi.nlm. nih.gov/ pubmed/22777751

[32]   Brama PA, Tekoppele JM, Bank RA, Karssenberg D, Barneveld A, van Weeren PR. Topographical mapping of biochemical properties of articular cartilage in the equine fetlock joint. Equine Vet J . 2000 Jan [cited 2017 Mar 21];32(1):19–26. Available from: http://www.ncbi.nlm.nih.gov/pubmed/10661380

[33]   Decker RS, Koyama E, Enomoto-Iwamoto M, Maye P, Rowe D, Zhu S, et al. Mouse limb skeletal growth and synovial joint development are coordinately enhanced by Kartogenin. Dev Biol . 2014 Nov 15 [cited 2017 Mar 21];395(2):255–67. Available from: http://www.ncbi.nlm.nih.gov/pubmed/25238962

[34]   Fujisawa T, Takigawa M, Kuboki T. [Cartilage and mechanical stress from the point of the view of development, growth, pathology and therapeutic aspects]. Clin

Calcium. 2004 Jul [cited 2017 Mar 8];14(7):29–35. Available from:
http://www.ncbi.nlm.nih.gov/ pubmed/15577073

[35]    Chung U, Kawaguchi H, Takato T, Nakamura K. Distinct osteogenic mechanisms
of bones of distinct origins. J Orthop Sci. 2004 Jul [cited 2017 Mar 21];9(4):410–4.
Available from: http://www.ncbi.nlm.nih.gov/ pubmed/15278782

[36]    Hoemann CD, Lafantaisie-Favreau C-H, Lascau-Coman V, Chen G, Guzmán-
Morales J. The cartilage-bone interface. J Knee Surg. 2012 May [cited 2017 Mar
8];25(2):85–97. Available from: http://www.ncbi.nlm.nih.gov/pubmed/22928426

[37]    Danfelter M, Onnerfjord P, Heinegard D. Fragmentation of Proteins in Cartilage
Treated with Interleukin-1: Specific cleavage of type IX collagen by matrix
metalloproteinase 13 releases the NC4 domain. J Biol Chem. 2007 Dec 21 [cited 2017
Mar 20];282(51):36933–41. Available from: http://www.ncbi.nlm.nih.gov/
pubmed/17951262

[38]    Heinegård D. Fell-Muir Lecture: Proteoglycans and more - from molecules to
biology. Int J Exp Pathol. 2009 Dec [cited 2017 Mar 20];90(6):575–86. Available from:
http://doi.wiley.com/10.1111/j.1365-2613.2009.00695.x

[39]    Farach-Carson MC, Hecht JT, Carson DD. Heparan Sulfate Proteoglycans: Key
Players in Cartilage Biology. Crit Rev Eukaryot Gene Expr. 2005 [cited 2017 Mar
8];15(1):29–48. Available from:
http://www.dl.begellhouse.com/journals/6dbf508d3b17c437,03d0336
f0e2ce096,115f023f2afbbf9b.htm

[40]    Watanabe H. [Cartilage proteoglycan aggregate: structure and function]. Clin
Calcium . 2004 Jul [cited 2017 Mar 20];14(7):9–14. Available from:
http://www.ncbi.nlm.nih.gov /pubmed/15577070

[41]    Kwon H, Paschos NK, Hu JC, Athanasiou K. Articular cartilage tissue
engineering: the role of signaling molecules. Cell Mol Life Sci. 2016 Mar 25 [cited
2017 Mar 8];73(6):1173–94. Available from:
http://www.ncbi.nlm.nih.gov/pubmed/26811234

[42]    Wilusz RE, Sanchez-Adams J, Guilak F. The structure and function of the
pericellular matrix of articular cartilage. Matrix Biol. 2014 Oct [cited 2017 Mar
8];39:25–32. Available from:
http://linkinghub.elsevier.com/retrieve/pii/S0945053X14001620

[43]    Tatari H. [The structure, physiology, and biomechanics of articular cartilage:
injury and repair]. Acta Orthop Traumatol Turc. 2007 [cited 2017 Mar 20];41 Suppl
2:1–5. Available from: http://www.ncbi.nlm.nih.gov/pubmed/18180577

[44]    Verdier M-P, Seite S, Guntzer K, Pujol J-P, Boumediene K. Immunohistochemical
analysis of transforming growth factor beta isoforms and their receptors in human
cartilage from normal and osteoarthritic femoral heads. Rheumatol Int. 2005 Mar 14
[cited 2017 Mar 8];25(2):118–24. Available from:
http://www.ncbi.nlm.nih.gov/pubmed/14618374

[45]   Nilsson O, Marino R, De Luca F, Phillip M, Baron J. Endocrine Regulation of the Growth Plate. Horm Res Paediatr. 2005 Nov 21 [cited 2017 Mar 8];64(4):157–65. Available from: http://www.ncbi.nlm.nih.gov/ pubmed/16205094.

[46]   Curl WW, Krome J, Gordon ES, Rushing J, Smith BP, Poehling GG. Cartilage injuries: a review of 31,516 knee arthroscopies. Arthroscopy 1997; 13(4):456–460.

[47]   Hjelle K, Solheim E, Strand T, Muri R, Brittberg M. Articular cartilage defects in 1,000 knee arthroscopies. Arthroscopy 2002;18(7):730–734.

[48]   Widuchowski W, Lukasik P, Kwiatkowski G, et al. Isolated full thickness chondral injuries. Prevalence and outcome of treatment. A retrospective study of 5233 knee arthroscopies. Acta Chir Orthop Traumatol Cech 2008;75(5):382–386.

[49]   Widuchowski W, Widuchowski J, Trzaska T. Articular cartilage defects: study of 25,124 knee arthroscopies. Knee. Jun 2007;14(3):177–82.

[50]   Widuchowski W, Widuchowski J, Faltus R, et al. Long-term clinical and radiological assessment of untreated severe cartilage damage in the knee: a natural history study. Scand J Med Sci Sports. Feb 2011;21(1):106–10.

[51]   Cicuttini F, Ding C, Wluka A, Davis S, Ebeling PR, Jones G. Association of cartilage defects with loss of knee cartilage in healthy, middle-age adults: a prospective study. Arthritis Rheum. Jul 2005;52(7):2033–9.

[52]   Davies-Tuck ML, Wluka AE, Wang Y, et al. The natural history of cartilage defects in people with knee osteoarthritis. Osteoarthritis Cartilage. Mar 2008;16(3):337–42.

[53]   Ding C, Cicuttini F, Scott F, Boon C, Jones G. Association of prevalent and incident knee cartilage defects with loss of tibial and patellar cartilage: a longitudinal study. Arthritis Rheum. Dec 2005;52(12):3918–27.

[54]   Messner K, Maletius W. The long-term prognosis for severe damage to weight-bearing cartilage in the knee: a 14-year clinical and radiographic follow-up in 28 young athletes. Acta Orthop Scand. Apr 1996;67(2):165–855.

[55]   Shelbourne KD, Jari S, Gray T. Outcome of untreated traumatic articular cartilage defects of the knee: a natural history study. J Bone Joint Surg Am. 2003;85(Suppl 2):8–16.

[56]   Mankin HJ. The response of articular cartilage to mechanical injury. The Journal of Bone & Joint Surgery. 1982;64(3):460–6.

[57]   Heinegard D, Oldberg A. 1989. Structure and biology of cartilage and bone matrix noncollagenous macromolecules. FASEB J. 1989 Jul;3(9):2042-51.

[58]   Shapiro F, Koide S, Glimcher MJ. Cell origin and differentiation in the repair of full-thickness defects of articular cartilage. The Journal of Bone & Joint Surgery. 1993;75(4):532–53.

[59]   Lewandowska K. Fibronectin-mediated adhesion of fibroblasts: inhibition by dermatan sulfate proteoglycan and evidence for a cryptic glycosaminoglycan- binding domain. The Journal of Cell Biology. 1987Jan;105(3):1443–54.

[60]    Hunziker E. Growth-factor-induced healing of partial-thickness defects in adult articular cartilage. Osteoarthritis and Cartilage. 2001;9(1):22–32.

[61]    Seeley MA, Knesek M, Vanderhave KL. Osteochondral Injury After Acute Patellar Dislocation in Children and Adolescents. Journal of Pediatric Orthopaedics. 2013;33(5):511–8.

[62]    Brophy RH, Zeltser D, Wright RW, Flanigan D. Anterior Cruciate Ligament Reconstruction and Concomitant Articular Cartilage Injury: Incidence and Treatment. Arthroscopy: The Journal of Arthroscopic & Related Surgery. 2010;26(1):112–20.

[63]    Nietosvaara YCACA, Aalto K, Kallio PE. Acute Patellar Dislocation in Children: Incidence and Associated Osteochondral Fractures. Journal of Pediatric Orthopaedics. 1994;14(4):513–5.

[64]    O'Donoghue DH. Chondral And Osteochondral Fractures. The Journal of Trauma: Injury, Infection, and Critical Care. 1966;6(4):469–81.

[65]    Benea H, Tomoaia G, Martin A, Bardas C. Arthroscopic management of proximal tibial fractures: technical note and case series presentation. Clujul Med. 2015;88(2):233-6. https://doi.org/10.15386/cjmed-422.

[66]    Faivre B, Benea H, Klouche S, Lespagnol F, Bauer T, Hardy P. An original arthroscopic fixation of adult's tibial eminence fractures using the Tightrope® device: a report of 8 cases and review of literature. Knee. 2014 Aug;21(4):833-9. https://doi.org/10.1016/j.knee.2014.02.007. Review.

[67]    Edmonds EW, Shea KG. Osteochondritis dissecans: editorial comment. Clin Orthop Relat Res 2013;471(4):1105–1106

[68]    Aichroth P. Osteochondritis dissecans of the knee: a clinical survey. J Bone Joint Surg Br. 1971;53(3):440–7.

[69]    Kocher MS. Management of Osteochondritis Dissecans of the Knee: Current Concepts Review. American Journal of Sports Medicine. 2006;34(7):1181–91.

[70]    Campbell CJ, Ranawat CS. Osteochondritis Dissecans. The Journal of Trauma: Injury, Infection, and Critical Care. 1966;6(2):201–21.

[71]    Gardiner TB. Osteochondritis dissecans in three members of one family. J Bone Joint Surg Br. 1955 Feb;37-B(1):139-41.

[72]    Barrie HJ. Hypertrophy and laminar calcification of cartilage in loose bodies as probable evidence of an ossification abnormality. J Pathol. 1980 Oct;132(2):161-8.

[73]    Green WT, Banks HH. Osteochondritis Dissecans in Children. The Journal of Bone & Joint Surgery. 1953;35(1):26–64.

[74]    Cahill BR. Osteochondritis Dissecans of the Knee: Treatment of Juvenile and Adult Forms. Journal of the American Academy of Orthopaedic Surgeons. 1995;3(4):237–47.

[75]    Cahill BR, Phillips MR, Navarro R. The results of conservative management of juvenile osteochondritis dissecans using joint scintigraphy. The American Journal of Sports Medicine. 1989;17(5):601–6.

[76]    ICRS SCORE/GRADE [Internet]. ICRS Main Site. [cited 2017Feb26]. Available from: http://cartilage.org/society/publications/icrs-score

[77]    Jose J, Pasquotti G, Smith MK, Gupta A, Lesniak BP, Kaplan LD. Subchondral insufficiency fractures of the knee: review of imaging findings. Acta Radiol. 2015 Jun;56(6):714-9. https://doi.org/10.1177/0284185114535132

[78]    Yamamoto T, Bullough PG. Spontaneous Osteonecrosis of the Knee: The Result of Subchondral Insufficiency Fracture. The Journal of Bone and Joint Surgery-American Volume. 2000;82(6):858–66.

[79]    Akamatsu Y, Mitsugi N, Hayashi T, Kobayashi H, Saito T. Low bone mineral density is associated with the onset of spontaneous osteonecrosis of the knee. Acta Orthopaedica. 2012;83(3):249–55.

[80]    Lotke PA, Nelson CL, Lonner JH. Spontaneous osteonecrosis of the knee: tibial plateaus. Orthopedic Clinics of North America. 2004;35(3):365–70.

[81]    Felson DT. Bone Marrow Edema and Its Relation to Progression of Knee Osteoarthritis. Annals of Internal Medicine. 2003Feb;139(5_Part_1):330.

[82]    Sharma L, Song J, Dunlop D, Felson D, Lewis CE, Segal N, et al. Varus and valgus alignment and incident and progressive knee osteoarthritis. Annals of the Rheumatic Diseases. 2010;69(11):1940–5.

[83]    Shelburne KB, Kim H-J, Sterett WI, Pandy MG. Effect of posterior tibial slope on knee biomechanics during functional activity. Journal of Orthopaedic Research. 2010;29(2):223–31.

[84]    Benea H. Associated procedures: instability, malalignment and meniscus lesions. Invited lecture at 1st SRATS Congress; 2016, Mar 31-Apr 1; Bucharest, Romania.

[85]    Cerejo R, Dunlop DD, Cahue S, Channin D, Song J, Sharma L. The influence of alignment on risk of knee osteoarthritis progression according to baseline stage of disease. Arthritis & Rheumatism. 2002;46(10):2632–6.

[86]    Brouwer GM, Tol AWV, Bergink AP, Belo JN, Bernsen RMD, Reijman M, et al. Association between valgus and varus alignment and the development and progression of radiographic osteoarthritis of the knee. Arthritis & Rheumatism. 2007;56(4):1204–11.

[87]    Neogi T, Zhang Y. Osteoarthritis prevention. Current Opinion in Rheumatology. 2011;23(2):185–91.

[88]    Tandogan RN, Taşer Ö, Kayaalp A, Taşkıran E, Pınar H, Alparslan B, et al. Analysis of meniscal and chondral lesions accompanying anterior cruciate ligament tears: relationship with age, time from injury, and level of sport. Knee Surgery, Sports Traumatology, Arthroscopy. 2003;12(4):262–70.

[89]    Seedhom BB. Transmission of the Load in the Knee Joint with Special Reference to the Role of the Menisci, PART I: anatomy, analysis and apparatus. Engineering in Medicine. 1979;8(4):207–19..

[90]    Lohmander LS, Englund PM, Dahl LL, Roos EM. The Long-term Consequence of Anterior Cruciate Ligament and Meniscus Injuries: Osteoarthritis. The American Journal of Sports Medicine. 2007;35(10):1756–69.

[91]    Jones HP, Appleyard RC, Mahajan S, Murrell GAC. Meniscal and Chondral Loss in the Anterior Cruciate Ligament Injured Knee. Sports Medicine. 2003;33(14):1075–89.

[92]    Chaudhari AMW, Briant PL, Bevill SL, Koo S, Andriacchi TP. Knee Kinematics, Cartilage Morphology, and Osteoarthritis after ACL Injury. Medicine & Science in Sports & Exercise. 2008;40(2):215–22.

[93]    Kessler MA, Behrend H, Henz S, Stutz G, Rukavina A, Kuster MS. Function, osteoarthritis and activity after ACL-rupture: 11 years follow-up results of conservative versus reconstructive treatment. Knee Surgery, Sports Traumatology, Arthroscopy. 2008;16(5):442–8.

[94]    Nakamae A, Engebretsen L, Bahr R, Krosshaug T, Ochi M. Natural history of bone bruises after acute knee injury: clinical outcome and histopathological findings. Knee Surgery Sports Traumatology Arthroscopy. 2006;14(12):1252–8.

[95]    Potter HG, Jain SK, Ma Y, Black BR, Fung S, Lyman S. Cartilage Injury After Acute, Isolated Anterior Cruciate Ligament Tear. The American Journal of Sports Medicine. 2012;40(2):276–85.

[96]    Dejour H, Bonnin M. Tibial translation after anterior cruciate ligament rupture. Two radiological tests compared. J Bone Joint Surg Br. 1994 Sep;76(5):745-9.

[97]    Kakarlapudi TK. Knee instability: isolated and complex. British Journal of Sports Medicine. 2000Jan;34(5).

[98]    Reilly DT, Martens M. Experimental Analysis of the Quadriceps Muscle Force and Patello-Femoral Joint Reaction Force for Various Activities. Acta Orthopaedica Scandinavica. 1972;43(2):126–37.

[99]    Garth WP. Clinical biomechanics of the patellofemoral joint. Operative Techniques in Sports Medicine. 2001;9(3):122–8.

[100]   Sullivan N, Robinson P, Ansari A, Hassaballa M, Robinson J, Porteous A, et al. Bristol index of patellar width to thickness (BIPWiT): A reproducible measure of patellar thickness from adult MRI. The Knee. 2014;21(6):1058–62.

[101]   Cetik O, Cift H, Comert B, Cirpar M. Risk of osteonecrosis of the femoral condyle after arthroscopic chondroplasty using radiofrequency: a prospective clinical series. Knee Surgery, Sports Traumatology, Arthroscopy. 2008;17(1):24–9.

[102]   Lansdown DA, Shaw J, Allen CR, Ma CB. Osteonecrosis of the Knee after Anterior Cruciate Ligament Reconstruction: A Report of 5 Cases. Orthop J Sports Med. 2015 Mar 24;3(3): 2325967115576120

[103]   Piper SL, Kramer JD, Kim HT, Feeley BT. Effects of local anesthetics on articular cartilage. Am J Sports Med. 2011 Oct;39(10):2245-53.

[104] Baker JF, Mulhall KJ. Local anesthetics and chondrotoxicty: What is the evidence? Knee Surg Sports Traumatol Arthrosc. 2012 Nov;20(11):2294-301.

[105] Wernecke C, Braun HJ, Dragoo JL. The Effect of Intra-articular Corticosteroids on Articular Cartilage. Orthopaedic Journal of Sports Medicine. 2015Apr;3(5):232596711558116.

[106] Sonnery-Cottet B, Archbold P, Thaunat M, Carnesecchi O, Tostes M, Chambat P. Rapid chondrolysis of the knee after partial lateral meniscectomy in professional athletes. The Knee. 2014;21(2):504–8.

[107] Steinmetz S, Bonnomet F, Rahme M, Adam P, Ehlinger M. Rapid chondrolysis of the medial knee compartment after arthroscopic meniscal resection: a case report. Journal of Medical Case Reports. 2016Jan;10(1).

[108] Buckwalter JA, Martin J, Mankin HJ. Synovial joint degeneration and the syndrome of osteoarthritis. Instr Course Lect. 2000;49:481-9.

[109] Buckwalter JA, Mankin HJ. Articular cartilage. II. Degeneration and osteoarthrosis, repair, regeneration and transplantation. J Bone Joint Surg Am 1997;79A: 612–32.

[110] Buckwalter JA, Martin J. Degenerative joint disease. Clin Symp. 1995;47(2):1-32.

[111] Smith KJ, Bertone AL, Weisbrode SE, Radmacher M. Gross, histologic, and gene expression characteristics of osteoarthritic articular cartilage of the metacarpal condyle of horses. American Journal of Veterinary Research. 2006;67(8):1299–306.

[112] Henrotin Y, Sanchez C, Balligand M. Pharmaceutical and nutraceutical management of canine osteoarthritis: Present and future perspectives. The Veterinary Journal. 2005;170(1):113–23.

[113] DePalma AA, Gruson KI. Management of cartilage defects in the shoulder. Curr Rev Musculoskeletal Med. 2012;5(3):254-262.

[114] Radsource. Osteochondral Injury of the Elbow [Internet]. WN Snearly. [cited 27 Februray 2017]. Available from: http://radsource.us/osteochondral-injury-of-the-elbow.

[115] Ho PC, Tse WI, Wong CWZ, Chow ECS. Arthroscopic Osteochondral Grafting for Radiocarpal Joint Defects. J Wrist Surg. 2013;2(3):212-219.

[116] Nieminen HJ, Zheng Y, Saarakkala S, Wang Q, Toyras J, Huang Y, Jurvelin J. Quantitative assessment of articular cartilage using high-frequency ultrasound: research findings and diagnostic prospects. Crit Rev Biomed Eng. 2009;37(6):461-94.

[117] Vanhoenacker FM, Maas M, Gielen JL. Imaging of Orthopaedic Sports Injuries. Berlin: Spinger-Verlag Berlin and Heidelberg GmbH & Co. KG. 2010.

[118] Nakasa T, Adachi N, Kato T, Ochi M. Appearance of Subchondral Bone in Computed Tomography Is Related to Cartilage Damage in Osteochondral Lesions of the Talar Dome. Foor Ankle Int. 2014;35(6):600-606.

[119] Kijowski R. Clinical Cartilage Imaging of the Knee and Hip Joints. American Journal of Roentgenology. 2010;195(3):618-628.

[120] Pope DT, Bloem HL, Beltran J, Morrison WB, Wilson DJ. Musculoskeletal Imaging. Elsevier Health Sciences. US. 2014

[121] Farr J, Gomoll AH. Cartilage Restoration. New York: Springer. 2014

[122] Rodrigues MB, Camanho GL. MRI Evaluation of Knee Cartilage. Revista Brasileira de Ortopedia. 2010;45(4):340-346.

[123] Gomoll AH, Yoshioka H, Watanabe A, Dunn JC, Minas T. Preoperative Measurement of Cartilage Defects by MRI Underestimates Lesion Size. Cartilage. 2011;2(4):389-393.

[124] Hepple S, Winson IG, Glew D: Osteochondral lesions of the talus: A revised classification. Foot Ankle Int. 1999;20:789-793.

[125] Battaglia M, Rimondi E, Monti C, Guaraldi F, Sant'Andrea A, Buda R, Cavallo M, Giannini S, Vannini F. Validity of T2 mapping in characterization of the regeneration tissue by bone marrow derived cell transplantation in osteochondral lesions of the ankle. Eur J Radiol. 2011;80(2):132-9.

[126] Matyat SJ, van Tiel J, Gold GE, Oei EHG. Quantitative MRI techniques of cartilage composition. Quant Imaging Med Surg. 2013;3(3):162-174.

[127] Louati K, Berenbaum F. Joint Biochemical Markers for Cartilage, Bone, Cartilage Degradation, Bone Remodeling, and Inflammation. IAPS. 2016.

[128] Siebuhr AS, He Y, Gudmann NS, Gram A, Kjelgaard-Petersen CF, Qvist P, Karsdal MA, Bay-Jensen AC. Biomarkers of cartilage and surrounding joint tissue. Biomarkers Med. 2014;8(5):713-731.

[129] Tomoaia G. Ortopedie. Ed. Medicală Iuliu Hațieganu, Cluj-Napoca, 2013. ISBN: 978-973-693-520-6.

[130] Veronesi F, Setti S, Buda R, Fini M. Pulsed electromagnetic fields combined with a collagenous scaffold and bone marrow concentrate enhance osteochondral regeneration: An in vivo study. Orthopedics and biomechanics. BMS Musculoskeletal Disorders. 2015;16:233.

[131] Kpk AC, van Bergen CJA, Tuijthof GJM, Klinkenbijl MN, van Noorden CJF, van Dijk CN, Kerkhoffs GMMJ. Macroscopic ICRS Poorly Correlates with O Driscoll Histological Cartilage Repair Assessment in a Goat Model. 2015;3:173.

[132] Zhang W, Lian Q, Li D, Wang K, Hao D, Bian W, He J, Jin Z. Cartilage Repair and Subchondral Bone Migration Using 3D Printing Osteochondral Composites: A One-Year-Period Study in Rabbit Trochlea. BioMed Research International. 2014(10):746138

[133] Falah M, Nierenberg G, Soudry M, Hayden M, Volpin G. Treatment of articular cartilage lesions of the knee. International Orthopaedics (SICOT). 2010;34:621-630.

[134] Jerosch J. Effects of Glucosamine and Chondroitin Sulfate on Cartilage Metabolism in OA: Outlook Nutrient Partners Especially Omega-3 Fatty Acids. 2011;2011;1-17.

[135]  van de Laar M, Pergolizzi Jr. JV, Mellinghoff HU, Merchante IM, Nalamachu S, O'Brien J, Perrot S, Raffa RB. Pain Treatment in Arthritis-Related Pain: Beyond NSAIDs. The Open Rheumatology Journal. 2012;6:320-330.

[136]  Paterno MV, Prokop TR, Schmitt LC. Physiscal Therapy Management of Patients with Osteochondritis Dissecans: A Comprensive Review. 2014;33:353-374.

[137]  Marinescu R, Laptoiu D, Antoniac I. Development of Modified Viscoelastic Solution with Magnetic Nanoparticles - Potential Method for Targeted Treatment of Chondral Injuries. Book Series: Key Engineering Materials. 2014;583:145-149.

[138]  Jo CH, Lee YG, Shin WH, Kim H, Chai JW, Jeong EC, Kim JE, Shin JS, Ra JC, Oh S, Yoon KS. Intra-articular injection of mesenchymal stem cells for the treatment of osteoarthritis of the knee: a proof-of-concept clinical trial. Stem Cells. 2014;32(5):1254-66.

[139]  Wernecke C, Braun HJ, Dragoo JL. The Effect of Intra-articular Corticosteroids on Articular Cartilage. A Systematic Review. 2015;3(5):1-7.

[140]  Redler LH, Cldwell JM, Schulz BM, Levine WN. Management of Articular Cartilage Defects of the Knee. The Physician and Sports Medicine. 2012;40(1):20-35.

[141]  Thiede RM, Markel MD. A review of the treatment methods for cartilage defects. Vet Comp Orthop Traumatol. 2012;25:263-272.

[142]  Gilogly SD, Newfield DM. Treatment of Articular Cartilage Defects of the Knee With Autologous Chondrocyte Implantation. Available from: http://www.medscape.com/ viewarticle/408519_2

[143]  van Oosterhout M, Sont JK, van Laar JM. Superior effect of arthroscopic lavage compared with needle aspiration in the treatment of inflammatory arthritis of the knee. Rheumatology. 2003;42:102-107.

[144]  Horton D, Anderson S, Hope NG. A review of current concepts in radiofrequency chondroplasty. ANZ Journal of Surgery. 2014;84:412-416.

[145]  Cole BJ, Pascual-Garrido C, Grumet RC. Surgical Management of Articular Cartilage Defects in the Knee. The Journal of Bone & Joint Surgery. 2009;91(7):1778-1790.

[146]  Trice ME, Bugbee WD, Greenwald AS, Phil D, Heim CS. Articular cartilage restoration: A review of currently available methods. American Academy of Orthopaedic Surgeons. 2010 77th Annual Meeting.

[147]  Mayo Clinic. Advances in Articular Cartilage Defect Management. Available from: http://www.mayoclinic.org/medical-professionals/clinical-updates/orthopedic-surgery/innovations-managing-articular-cartilage-defects-knee

[148]  Richter DL, Schenck Jr RC, Wascher DC, Treme G. Knee Articular Cartilage Repair and Restoration Techniques: A Review of the Literature. Sports Health. 2016 Mar-Apr;8(2):153-60.

[149]  Hangody L, Ráthonyi GK, Duska Z, Vásárhelyi G, Füles P, Módis L. Autologous Osteochondral Mozaicplasty. J Bone Joint Surg Am. 2004;86(1):65-72.

[150] Marcacci M, Kon E, Delcogliano M, Filardo G, Busacca M, Zaffagnini S. Arthroscopic autologous osteochondral grafting for cartilage defects of the knee: prospective study results at a minimum 7-year follow-up. Am J Sports Med. 2007;35(12):2014-21.

[151] Bostrom MPG, Seigerman DA. The Clinical Use of Allografts, Demineralized Bone Matrices, Synthetic Bone Graft Substitutes and Osteoinductive Growth Factors: A Survey Study. 2005;1(1):9-18.

[152] Aubin PP, Cheah HK, Davis AM, Gross AE. Long term follow-up of fresh femoral osteochondral allografts for post traumatic defects of the knee. Clin Orthop Relat Res. 2001;391S:318-27.

[153] Toolan BC, Frenkel SR, Pereira DS, Alexandre H. Development of a novel osteochondral graft for cartilage repair. J Biomed Mater Res. 1998;41(2):244-50.

[154] DiDomenico LA, Thomas ZM. Osteobiologics in Foot and Ankle Surgery. Clin Podiatr Med Surg. 2015;32:1-19.

[155] Lee SK. Fractures of the carpal bones. Chapter in Wolfe SW, Pederson WC, Hotchkiss RN, Kozin SH, Cohen MS (Eds.): Green's Operative Hand Surgery, 7th Edition, New York, Elsevier Health Sciences, Feb. 2016.

[156] Rau JV, Antoniac I, Cama G, Komlev VS, Ravaglioli A. Bioactive Materials for Bone Tissue Engineering. BioMed research international 2016. Article Number: 3741428.

[157] Laursen JO. Treatment of full-thickness cartilage lesions and early OA using large resurfacing prosthesis: UniCAP. Knee Surg Sports Traumatol Arthrosc. 2016;24(5):1695-701.

[158] Seo SS, Kim CW, Jung DW. Management of Focal Chondral Lesion in the Knee Joint. Knee Surg Relat Res. 2011;23(4):185-196.

[159] Gobbi A, Bathan L. Biological approaches for cartilage repair. J Knee Surg. 2009 Jan;22(1):36-44.

[160] Rau JV, Antoniac I, Fosca M, et al. Glass-ceramic coated Mg-Ca alloys for biomedical implant applications. Materials Science & Engineering C-Materials for Biological Applications. 2016;64:362-369.

[161] Petreus T, Stoica BA, Petreus O, Goriuc A, Cotrutz CE, Antoniac IV. Preparation and cytocompatibility evaluation for hydrosoluble phosphorous acid-derivatized cellulose as tissue engineering scaffold material. Journal of Materials Science: Materials in Medicine, 2014;25(4):1115-1127.

[162] Gobbi A, Kumar A, Karnatzikos G, Nakamura N. The Future of Cartilage Repair Surgery. Chapter in Shetty AA et al (eds.): Techniques in Cartilage Repair Surgery. Berlin, Heidelberg: Springer Verlag; Apr 2014. Pp DOI 10.1007/978-3-642-41921-8_31, © ESSKA 2014.

[163] Enea D, Cecconi S, Busilacchi A, Manzotti S, Gesuita R, Gigante A. Matrix-induced autologous chondrocyte implantation (MACI) in the knee. Knee Surg Sports Traumatol Arthrosc. 2012;20:862-869.

[164] Nixon AJ, Rickey E, Butler TJ, Scimeca MS, Moran N, Matthews GL. A chondrocyte infiltrated collagen type I/III membrane (MACI® implant) improves cartilage healing in the equine patellofemoral joint model. Osteoarthritis Cartilage. 2015 Apr;23(4):648-60. https://doi.org/10.1016/j.joca.2014.12.021. Epub 2015 Jan 7.

[165] Özmeriç A, Alemdaroğlu KB, Azdoğan NH. Treatment for cartilage injuries of the knee with a new treatment algorithm. World J Orthop. 2014;5(5):677-684.

[166] Ullah I, Subbarao RB, Rho GJ. Human mesenchymal stem cells – current trends and future prospective. Biosci Rep. 2015;35(2);e00191.

[167] Pastides P, Chimutengwende-Gordon M, Maffulli N, Khan W. Stem cell therapy for human cartilage defects: a systematic review. Osteoarthritis and Cartilage (2013), http://dx.doi.org/10.1016/j.joca.2013.02.008

[168] Muhammad H, Schminke B, Miosge N. Current concepts in stem cell therapy for articular cartilage repair. Expert Opin Biol Ther. 2013;13(4):541-548.

[169] Buda R, Vannini F, Cavallo M, Grigolo B, Cenacchi A, Giannini S. Osteochondral lesions of the knee: a new one-step repair technique with bone-marrow-derived cells. J Bone Joint Surg Am. 2010;92:2-11.

[170] Fortier LA, Potter HG, Rickey EJ, Schnabel LV, Foo LF, Chong LR, et al. Concentrated bone marrow aspirate improves full-thickness cartilage repair compared with microfracture in the equine model. J Bone Joint Surg Am. 2010 Aug 18;92(10):1927-37.

[171] Gobbi A, Karnatzikos G, Scotti C, Mahajan V, Mazzucco L, Grigolo B. One-Step Cartilage Repair with Bone Marrow Aspirate Concentrated Cells and Collagen Matrix in Full-Thickness Knee Cartilage Lesions: Results at 2-Year Follow-up. Cartilage. 2011;2:286-299.

[172] Kobayashi T, Ochi M, Yanada S et al. A novel cell delivery system using magnetically labeled mesenchymal stem cells and an external magnetic device for clinical cartilage repair. Arthroscopy. 2008;24(1):69-76.

[173] Benea H, Muste A, Berce C, Soritau O, Pasca RD, Tomoaia G. One-step reconstruction of articular cartilage using collagen scaffolds and autologous cells – our experience. Oral communication at BIOMMEDD International Conference, Sept. 15-17, 2016, Constanta, Romania.

[174] Gao J, Yao JQ, Caplan AI. Stem cells for tissue engineering of articular cartilage. Proc Inst Mech Eng H. 2007 Jul;221(5): 441-50. Review.

[175] Benea H, Tomoaia G, Soritau O, Pasca RD. A review on the reconstruction of articular cartilage using collagen scaffolds. Rom Biotech Letters. 2016;21(4):11735-11742.

[176] Benea H, d'Astorg H, Klouche S, Bauer T, Tomoaia G, Hardy P. Pain evaluation after all-inside anterior cruciate ligament reconstruction and short term functional results of a prospective randomized study. The Knee. 2014;21(1):102–106. ISSN: 0968-0160.

[177] Loia MC, Vanni S, Rosso F, Bonasia DE, Bruzzone M, Dettoni F, Rossi R. High tibial osteotomy in varus knees: indications and limits. Joints. 2016;4(2):98-110.

[178] Botez P, Sirbu P, Simion L, Antoniac I, et al. Application of a biphasic macroporous synthetic bone substitutes CERAFORM (A (R)): clinical and histological results. EJOST. 2009;19(6):387-395.

[179] Koshino T, Wada S, Ara Y, Saito T. Regeneration of degenerated articular cartilage after high tibial valgus osteotomy for medial compartmental osteoarthritis of the knee. Knee. 2003;10(3):229-36.

[180] Lee DC, Byun SJ. High Tibial Osteotomy. Knee Surg Relat Res. 2012;24(2):61-69.

[181] an CG, Tang FW, Zhang QJ, Lu SH, Liu HY, Zhao ZM et al. Characterization and neural differentiation of fetal lung mesenchymal stem cells. Cell Transplant. 2005;14(5):311-21.

[182] Gronthos S, Mankani M, Brahim J, Robey PG, Shi S. Postnatal human dental pulp stem cells (DPSCs) in vitro and in vivo. Proc Natl Acad Sci U S A. 2000 Dec 5;97(25):13625-30.

[183] Karaöz E, Doğan BN, Aksoy A, Gacar G, Akyüz S, Ayhan S et al. Isolation and in vitro characterization of dental pulp stem cells from natal teeth. Histochem Cell Biol. 2010 Jan;133(1):95-112. https://doi.org/10.1007/s00418-009-0646-5.

[184] Sonoyama W, Liu Y, Fang D, Yamaza T, Seo BM, Zhang C et al. Mesenchymal stem cell-mediated functional tooth regeneration in swine. PLoS One. 2006 Dec 20;1:e79.

[185] Caballero M, Reed CR, Madan G, van Aalst JA. Osteoinduction in umbilical cord- and palate periosteum-derived mesenchymal stem cells. Ann Plast Surg. 2010 May;64(5):605-9. https://doi.org/10.1097/SAP.0b013e3181ce3929.

[186] Seo BM, Miura M, Gronthos S, Bartold PM, Batouli S, Brahim J et al. Investigation of multipotent postnatal stem cells from human periodontal ligament. Lancet. 2004 Jul 10-16;364(9429):149-55.

[187] Park JY, Jeon SH, Choung PH. Efficacy of periodontal stem cell transplantation in the treatment of advanced periodontitis. Cell Transplant. 2011;20(2):271-85. https://doi.org/10.3727/096368910X519292.

[188] Nayernia K, Lee JH, Drusenheimer N, Nolte J, Wulf G, Dressel R et al. Derivation of male germ cells from bone marrow stem cells. Lab Invest. 2006 Jul;86(7):654-63. Epub 2006 May 1.

[189] Huselstein C, Li Y, He X. Mesenchymal stem cells for cartilage engineering. Biomed Mater Eng. 2012;22(1-3):69-80. https://doi.org/10.3233/BME-2012-0691.

[190] Leatherman J. Stem cells supporting other stem cells. Front Genet. 2013 Dec 3;4:257. https://doi.org/10.3389/fgene.2013.00257.

[191] Pfister O, Della Verde G, Liao R, Kuster GM. Regenerative therapy for cardiovascular disease. Transl Res. 2014 Apr;163(4):307-20. https://doi.org/10.1016/j.trsl.2013.12.005. Epub 2013 Dec 11.

[192] Fadel L, Viana BR, Feitosa ML, Ercolin AC, Roballo KC, Casals JB et al. Protocols for obtainment and isolation of two mesenchymal stem cell sources in sheep. Acta Cir Bras. 2011 Aug;26(4):267-73.

[193] Liu Z, Wang W, Gao J, Zhou H, Zhang Y. Isolation, culture, and induced multiple differentiation of Mongolian sheep bone marrow-derived mesenchymal stem cells. In Vitro Cell Dev Biol Anim. 2014;50(5):464-74. https://doi.org/10.1007/s11626-013-9725-y.

[194] Odabas S, Elçin AE, Elçin YM. Isolation and characterization of mesenchymal stem cells.. Methods Mol Biol. 2014;1109:47-63.

[195] Galateanu B, Dinescu S, Cimpean A, Dinischiotu A, Costache M. Modulation of Adipogenic Conditions for Prospective Use of hADSCs in Adipose Tissue Engineering. Int. J. Mol. Sci. 2012, 13, 15881-15900; doi:10.3390/ijms131215881

[196] Lupescu O, Antoniac I. Bone Substitutes in Orthopedic and Trauma Surgery. Bioceramics and Biocomposites: From Research to Clinical Practice. 2019:341-366.

[197] Antoniac I, Sinescu C, Antoniac A. Adhesion aspects in biomaterials and medical devices. Journal of Adhesion Science and Technology. 2016 Aug;30(16):1711-1715.

[198] Antoniac I, Vranceanu MD, Antoniac A. The influence of the magnesium powder used as reinforcement material on the properties of some collagen based composite biomaterials. Journal of Optoelectronics and Advanced Materials. 2013 Jul;15(7-8):667-672.

[199] Kohli N, Wright KT, Sammons RL, Jeys L, Snow M, Johnson WE. An In Vitro Comparison of the Incorporation, Growth, and Chondrogenic Potential of Human Bone Marrow versus Adipose Tissue Mesenchymal Stem Cells in Clinically Relevant Cell Scaffolds Used for Cartilage Repair. Cartilage. 2015 Oct;6(4):252-63. https://doi.org/10.1177/1947603515589650.

[200] Vranceanu MD, Antoniac I, Miculescu F, et al. The influence of the ceramic phase on the porosity of some biocomposites with collagen matrix used as bone substitutes. Journal of Optoelectronics and Advanced Materials. 2012 Jul;14(7-8):671-677.

[201] Miculescu F, Bojin D, Ciocan, LT, Antoniac I, et al. Experimental researches on biomaterial-tissue interface interactions. Journal of Optoelectronics and Advanced Materials. 2007;9(11):3303-3306.

[202] Antoniac I. Biodegradability of some collagen sponges reinforced with different bioceramics. Book Series: Key Engineering Materials. 2014;587:179-184.

[203] Bowen RE. Stromal Vascular Fraction from Lipoaspirate Infranatant: Comparison between Suction-Assisted Liposuction and Nutational Infrasonic Liposuction. Aesthetic Plast Surg. 2016 Jun;40(3):367-71. https://doi.org/10.1007/s00266-016-0631-z.

[204] Grässel S, Lorenz J. Tissue-engineering strategies to repair chondral and osteochondral tissue in osteoarthritis: use of mesenchymal stem cells. Curr Rheumatol Rep. 2014 Oct;16(10):452. https://doi.org/10.1007/s11926-014-0452-5. Review. PubMed PMID: 25182680; PubMed Central PMCID: PMC4182613.

[205] Fellows CR, Matta C, Zakany R, Khan IM, Mobasheri A. Adipose, Bone Marrow and Synovial Joint-Derived Mesenchymal Stem Cells for Cartilage Repair. Front Genet. 2016 Dec 20;7:213. https://doi.org/10.3389/fgene.2016.00213. Review. PubMed PMID: 28066501; PubMed Central PMCID: PMC5167763.

[206] Wei Y, Sun X, Wang W, Hu Y. Adipose-derived stem cells and chondrogenesis. Cytotherapy. 2007;9(8):712-6. Epub 2007 Oct 4. Review.

[207] Wu L, Cai X, Zhang S, Karperien M, Lin Y. Regeneration of articular cartilage by adipose tissue derived mesenchymal stem cells: perspectives from stem cell biology and molecular medicine. J Cell Physiol. 2013 May;228(5):938-44. https://doi.org/10.1002/jcp.24255. Review.

[208] Barba M, Cicione C, Bernardini C, Michetti F, Lattanzi W. Adipose-derived mesenchymal cells for bone regeneration: state of the art. Biomed Res Int. 2013;2013:416391. https://doi.org/10.1155/2013/416391.

[209] Cicione C, Di Taranto G, Barba M, Isgrò MA, D'Alessio A, Cervelli D, Lattanzi W et al. In Vitro Validation of a Closed Device Enabling the Purification of the Fluid Portion of Liposuction Aspirates. Plast Reconstr Surg. 2016 Apr;137(4):1157-67. https://doi.org/10.1097/ PRS.0000000000002014.

[210] Yoshimura K, Shigeura T, Matsumoto D, Sato T, Takaki Y, Aiba-Kojima E, et al. Characterization of freshly isolated and cultured cells derived from the fatty and fluid portions of liposuction aspirates. J Cell Physiol. 2006 Jul;208(1):64-76.

[211] Wayne JS, McDowell CL, Shields KJ, Tuan RS. In vivo response of polylactic acid-alginate scaffolds and bone marrow-derived cells for cartilage tissue engineering. AP Tissue Eng. 2005 May-Jun;11(5-6):953-63.

[212] Saha S, Kirkham J, Wood D, Curran S, Yang XB. Informing future cartilage repair strategies: a comparative study of three different human cell types for cartilage tissue engineering. Cell Tissue Res. 2013 Jun;352(3):495-507. https://doi.org/10.1007/s00441-013-1586-x.

[213] Sellin Jeffries MK, Kiss AJ, Smith AW, Oris JT. A comparison of commercially-available automated and manual extraction kits for the isolation of total RNA from small tissue samples. BMC Biotechnol. 2014 Nov 14;14:94. https://doi.org/10.1186/s12896-014-0094-8.

[214] Saulnier N, Piscaglia AC, Puglisi MA, Barba M, Arena V, Pani G et al. Molecular mechanisms underlying human adipose tissue-derived stromal cells differentiation into

a hepatocyte-like phenotype. Dig Liver Dis. 2010 Dec;42(12):895-901. https://doi.org/10.1016/j.dld.2010.04.013.

[215] Livak KJ, Schmittgen TD. Analysis of relative gene expression data using real-time quantitative PCR and the 2(-Delta Delta C(T)) Method. Methods. 2001 Dec;25(4):402-8.

[216] Zhu M, Gao JH, Lu F. Cell biological study of cultured cells derived from the fatty of fluid portions of liposuction aspirates]. Zhonghua Zheng Xing Wai Ke Za Zhi. 2008 Mar;24(2):138-44.

[217] Arpad S, Trambitas C, Matei E, Antoniac I, et al. Effect of Osteoplasty with Bioactive Glass (S53P4) in Bone Healing - In vivo Experiment on Common European Rabbits (Oryctolagus cuniculus). Revista de Chimie. 2018;69(2):429-433.

[218] Dong Z, Luo L, Liao Y, Zhang Y, Gao J, Ogawa R, Ou C, Zhu M, Yang B, Lu F. In vivo injectable human adipose tissue regeneration by adipose-derived stem cells isolated from the fluid portion of liposuction aspirates. Tissue Cell. 2014 Jun;46(3):178-84. https://doi.org/10.1016/j.tice.2014.04.001.

[219] Anders S, Volz M, Frick H, Gellissen J. A Randomized, Controlled Trial Comparing Autologous Matrix-Induced Chondrogenesis (AMIC®) to Microfracture: Analysis of 1- and 2-Year Follow-Up Data of 2 Centers. Open Orthop J. 2013 May 3;7:133-43. https://doi.org/10.2174/1874325001307010133.

[220] Smith P, Adams WP Jr, Lipschitz AH, Chau B, Sorokin E, Rohrich RJ et al. Autologous human fat grafting: effect of harvesting and preparation techniques on adipocyte graft survival. Plast Reconstr Surg. 2006 May;117(6):1836-44.

[221] Veronesi F, Cadossi M, Giavaresi G, Martini L, Setti S, Buda R, et al. Pulsed electromagnetic fields combined with a collagenous scaffold and bone marrow concentrate enhance osteochondral regeneration: an in vivo study. BMC Musculoskeletal Disorders (2015) 16:233. DOI 10.1186/s12891-015-0683-2.

[222] Centrul National de Rezonata Magnetica: http://phys.ubbcluj.ro/laboratoare/cnrm/en-noutati.html

[223] Hennig J, Nauerth A, Friedburg H. RARE imaging: a fast imaging method for clinical MR., Magn. Reson. Med., vol. 3, no. 6, pp. 823–833, 1986.

[224] Haase A. Snapshot Flash MRI Applications to T1,Tt2 and chemical-shift imaging, Magn. Reson. Med., vol. 13, no. 1, pp. 77–89, 1990.

[225] Kon E, Filardo G, Di Matteo B, Perdisa F, Marcacci M. Matrix assisted autologous chondrocyte transplantation for cartilage treatment: A systematic review. Bone & joint research. 2013;2(2):18-25.

[226] Rai V, Dilisio MF, Dietz NE, Agrawal DK. Recent Strategies in Cartilage Repair: A Systemic Review of the Scaffold Development and Tissue Engineering. J Biomed Mater Res A. 2017 Apr 7. https://doi.org/10.1002/jbm.a.36087. [Epub ahead of print]

[227] Vidal MA, Robinson SO, Lopez MJ, Paulsen DB, Borkhsenious O, Johnson JR, et al. Comparison of chondrogenic potential in equine mesenchymal stromal cells derived

from adipose tissue and bone marrow. Vet Surg. 2008 Dec;37(8):713-24.
https://doi.org/10.1111/j.1532-950X.2008.00462.x.

[228] Hui JH, Chen F, Thambyah A, Lee EH. Treatment of chondral lesions in advanced osteochondritis dissecans: a comparative study of the efficacy of chondrocytes, mesenchymal stem cells, periosteal graft and mozaicplasty (osteochondral autograft) in animal models. J Pediatri Orthop. 2004;24:427-33.

[229] Kim SH, Park DY, Min BH. A New Era of Cartilage Repair using Cell Therapy and Tissue Engineering: Turning Current Clinical Limitations into New Ideas. Tissue Engineering and Regenerative Medicine. 2012;9(5):240-248. DOI 10.1007/s13770-012-0370-4.

[230] Nakamura T, Sekiya I, Muneta T et al. Arthroscopic, histological and MRI analyses of cartilage repair after a minimally invasive method of transplantation of allogeneic synovial mesenchymal stromal cells into cartilage defects in pigs. Cytotherapy. 2012 Mar;14(3):327-38.

[231] Gomoll AH, Filardo G, de Girolamo L, Espregueira-Mendes J, Marcacci M, Rodkey WG, Steadman JR, et al. Surgical treatment for early osteoarthritis. Part I: cartilage repair procedures. Knee Surgery Sports Traumatology Arthroscopy. 2011;20(3):450-66.

[232] Manda K, Ryd L, Eriksson A, Finite element simulations of a focal knee resurfacing implant applied to localized cartilage defects in a sheep model, J Biomech. 2011 Mar 15;44(5):794-801. https://doi.org/10.1016/j.jbiomech.2010.12.026.

[233] Taylor WR, Poepplau BM, König C, Ehrig RM, Zachow S, Duda GN, Heller MO., The medial-lateral force distribution in the ovine stifle joint during walking, J Orthop Res. 2011 Apr;29(4):567-71. https://doi.org/10.1002/jor.21254. Epub 2010 Oct 18

[234] Bahraminasaba m, et al., Finite Element Analysis of the Effect of Shape Memory Alloy on the Stress Distribution and Contact Pressure in Total Knee Replacement, Trends Biomater. Artif. Organs, 25(3), 95-100 (2011).

[235] Venäläinen MS, Mononen ME, Salo J, Räsänen LP, Jurvelin JS, Töyräs J, Virén T, Korhonen RK, Quantitative Evaluation of the Mechanical Risks Caused by Focal Cartilage Defects in the Knee, Sci Rep. 2016 Nov 29;6:37538. https://doi.org/10.1038/srep37538.

[236] Cosma C, Cercetări privind îmbunătățirea fabricației implanturilor medicale din titan, prin topire selectivă cu laser, Teza de doctorat, Universitatea Tehnica Cluj-Napoca, 2015.

[237] Murr L.E., Gaytan S.M., Martinez E, Medina F, Wicker R.B., Next Generation Orthopaedic Implants by Additive Manufacturing Using Electron Beam Melting, International Journal of Biomaterials, Article ID 245727, DOI:10.1155/2012/245727, 2012.

[238] Nakano T, Ishimoto T, Powder-based Additive Manufacturing for development of Tailor-made implants for orthopedic applications, KONA Powder and Particle Journal, Vol. 32, Pp.75-84, HTTPS://DOI.ORG/http://doi.org/10.14356/kona.2015015, 2015.

[239] Imanishi J, Choong P.F., Three-dimensional printed calcaneal prosthesis following total calcanectomy, Int J Surg Case Rep, 10: 83–87, 2015.

[240] Kunz M, Devlin S, Gong RH, Inoue J, Waldman SD, Hurtig M, Abolmaesumi P, Stewart J., Prediction of the repair surface over cartilage defects: a comparison of three methods in a sheep model, Med Image Comput Assist Interv. 2009;12(Pt 1):75-82.

[241] Armstrong SJ, Read RA, Price R., Topographical variation within the articular cartilage and subchondral bone of the normal ovine knee joint: a histological approach, Osteoarthritis Cartilage. 1995 Mar;3(1):25-33.

[242] Knox P, Levick JR, McDonald JN, Synovial fluid--its mass, macromolecular content and pressure in major limb joints of the rabbit, Q J Exp Physiol. 1988 Jan;73(1):33-45.

[243] Moran CJ, Ramesh A, Brama PA, O'Byrne JM, O'Brien FJ, Levingstone TJ, The benefits and limitations of animal models for translational research in cartilage repair, J Exp Orthop. 2016 Dec;3(1):1. https://doi.org/10.1186/s40634-015-0037-x. Epub 2016 Jan 6.

[244] Pearce AI, Richards RG, Milz S, Schneider E, Pearce SG., Animal models for implant biomaterial research in bone: a review, Eur Cell Mater. 2007 Mar 2;13:1-10.

[245] Zhao Y, Wang W, Xin H, Zang S, Zhang Z, Wu Y,The remodeling of alveolar bone supporting the mandibular first molar with different levels of periodontal attachment, Med Biol Eng Comput, 2013 Sep;51(9):991-7.

[246] Tanaka K., Tanimoto Y, Kita Y, Enoki S, Katayama T, The Effects of Trabecular Bone Microstructure on Compression Property of Bovine Cancellous Bone, Key Engineering Materials, Vols. 452-453, pp. 297-300, 2011.

[247] Pal S, Design of Artificial Human Joints & Organs, Chapter 2, Mechanical Properties of Biological Materials, Springer, 2014, DOI 10.1007/978-1-4614-6255-2_2.

[248] Szabelska A, Tatara MR, Krupski W., Morphological, densitometric and mechanical properties of mandible in 5-month-old Polish Merino sheep, BMC Vet Res. 2017 Jan 5;13(1):12. https://doi.org/10.1186/s12917-016-0921-3.

[249] Homicz MR, McGowan KB, Lottman LM, Beh G, Sah RL, Watson D, A compositional analysis of human nasal septal cartilage, Arch Facial Plast Surg., 2003 Jan-Feb;5(1):53-8.

[250] Wang Y, Fan Y, Zhang M, Comparison of stress on knee cartilage during kneeling and standing using finite element models, Med Eng Phys. 2014 Apr;36(4):439-47. https://doi.org/10.1016/j.medengphy.2014.01.004. Epub 2014 Feb 5.

[251] Mansour JM. Biomechanics of cartilage. In Kinesiology: The Mechanics and Pathomechanics of Human Movement: Second Edition 2013 Jul 5. Wolters Kluwer Health.

[252] Vickers SM, Gotterbarm T, Spector M., Cross-linking affects cellular condensation and chondrogenesis in type II collagen-GAG scaffolds seeded with bone

marrow-derived mesenchymal stem cells, J Orthop Res. 2010 Sep;28(9):1184-92. https://doi.org/10.1002/jor.21113.

[253] Bandyopadhyay-Ghosh S., Bone as a Collagen-hydroxyapatite Composite and its Repair, Trends Biomater. Artif. Organs, Vol 22(2), 2008.

[254] Matthew D. Shoulders, Ronald T. Raines, Collagen Structure and Stability, Ann Rev Biochem. 2009; 78: 929–958.

[255] Barkaoui A, Hambli R., Nanomechanical properties of mineralized collagen microfibrils based on finite elements method: biomechanical role of cross-links, Comput Methods Biomech Biomed Engin. 2014;17(14):1590-601. https://doi.org/10.1080/10255842.2012.758255.

[256] Wenger MPE, Bozec L, Horton MA, Mesquida P, Mechanical Properties of Collagen Fibrils, Biophysical Journal Volume 93 August 2007 1255–1263.

[257] Brand RA, Joint contact stress: a reasonable surrogate for biological processes? Iowa Orthop J. 2005;25:82-94.

[258] Miculescu F, Jepu I, Lungu CP, Antoniac I, et al. Correlation Between Simulation and Microanalytical Experiment in Multilayer Nanostructures Type Analysis. UPB Scientific Bulletin - Series A – Applied Mathematics and Physics. 2011;73(4):153-166.

## Keyword index

**About the author**

*Lect.* ***Horea Rares Ciprian BENEA***, *MD, PhD,* is a consultant in Orthopedics and Traumatology surgery with 14 years of experience and since 2019 he is the head of University Clinic of Orthopedics and Traumatology from Cluj-Napoca, Romania. He takes care of a great variety of osteo-articular traumatic and non-traumatic pathology, but his focus domains are: arthroscopy (shoulder and knee), joint replacement (shoulder, hip, knee), cartilage reconstruction surgery and autologous biological therapies (stem cells, PRP).

He holds a PhD in Medicine from the University of Medicine and Pharmacy "Iuliu Hatieganu" Cluj-Napoca. Since 2017 he is Lecturer at the Orthopedics and Traumatology Discipline from UMF "Iuliu Hatieganu" Cluj-Napoca, where he teaches the French, English and Romanian series of students. He is the founder of CBSO – Center of Biomedical Sciences applied in Orthopedics, an integrated clinical and academic center for multidisciplinary research, under the frame of the Medical University. Lect. Horea BENEA, MD PhD, has published widely more than 40 peer-reviewed papers, proceedings and abstracts, very appreciated and cited in the literature, and presented more than 35 oral communications or poster at international conferences. He has also experience in research projects as leader or team member. The publications cover a wide area of subjects, from clinical practice studies to top research in the domain of regenerative medicine, implant tolerance and osseointegration and osteochondral reconstruction by tissue engineering and autologous biological means, including here multipotent stem cells and collagen membranes. Apart this, the present area of research extended towards the domain of regenerative medicine, focusing on osteoarthritis treatment. His research activity was awarded by the Faculty of Medicine of UMF "Iuliu Hatieganu" Cluj-Napoca with the Award for Young Surgeons in 2013 and other three awards for communications presented.

His international networking includes membership in the Communication Workgroup of European Shoulder Associates (ESA) under the frame of European Society of Sports Traumatology, Knee Surgery and Arthroscopy (ESSKA), an active association for professionals of arthroscopy, sports medicine and reconstructive surgery. He is member of Scientific Committee and General Assembly of the Romania Society of Sports Trauma and Arthroscopy (SRATS).

## ANNEX 8.1

CT images of in vivo tests results at 7 months after treatment with collagen membrane alone (Case 1)

| *Fig. 1.* Initial section of the treated zone (42.81), coronal view | *Fig. 2.* Section 46.09 in regard with treated zone, coronal view | *Fig. 3.* Section 48.13 in regard with treated zone, coronal view | *Fig. 4.* Final section 50.00 in regard with treated zone, coronal view |
|---|---|---|---|

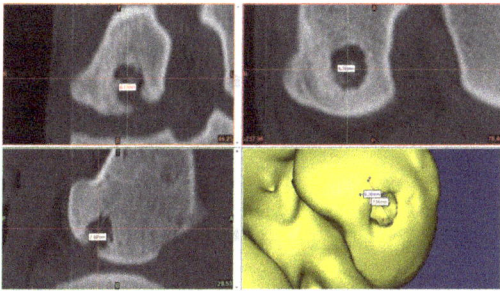

*Fig. 5.* CT images of the zones where bone tissue did not develop at 7 months postoperatively (coronal, axial, sagittal, 3D render)

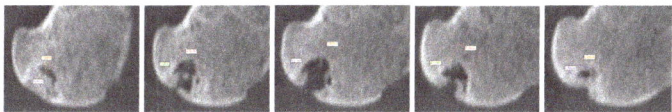

*Fig. 6.* Following the defect: first section 27.03, last section 32.24, sagittal view

## ANNEX 8.2

CT images of in vivo tests results at 7 months after treatment with collagen membrane and bone marrow concentrate (Case 2)

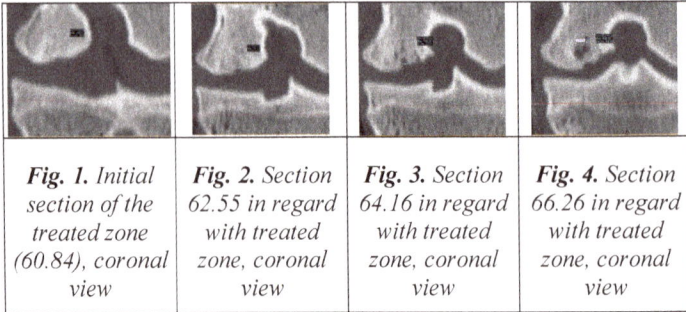

| ***Fig. 1.*** *Initial section of the treated zone (60.84), coronal view* | ***Fig. 2.*** *Section 62.55 in regard with treated zone, coronal view* | ***Fig. 3.*** *Section 64.16 in regard with treated zone, coronal view* | ***Fig. 4.*** *Section 66.26 in regard with treated zone, coronal view* |
|---|---|---|---|

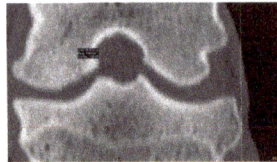

***Fig. 5.*** *Final section 68.11 in regard with treated zone, coronal view*

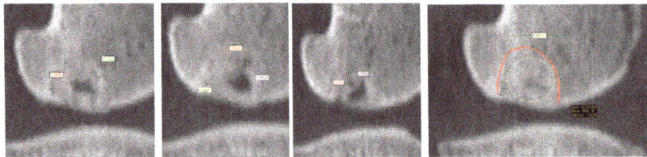

***Fig. 6.*** *Following the defect: first section 18.85, last section 20.80, sagittal view*

**ANNEX 8.3**

CT images of in vivo tests results at 7 months after treatment with collagen membrane
and adipose derived stem cells (Case 3)

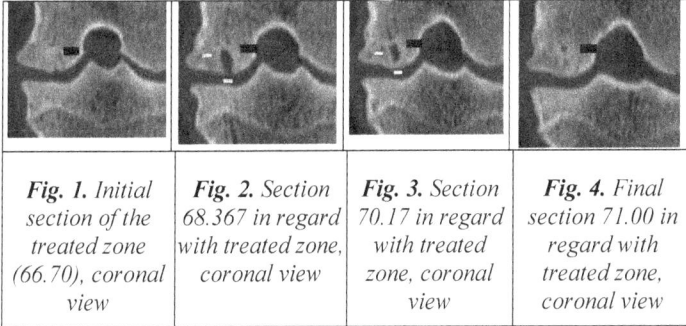

| **Fig. 1.** *Initial section of the treated zone (66.70), coronal view* | **Fig. 2.** *Section 68.367 in regard with treated zone, coronal view* | **Fig. 3.** *Section 70.17 in regard with treated zone, coronal view* | **Fig. 4.** *Final section 71.00 in regard with treated zone, coronal view* |
|---|---|---|---|

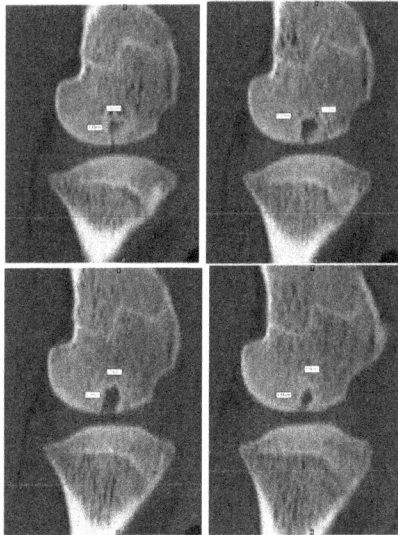

**Fig. 5.** *Following the defect: first section 17.20, last section 19.69, sagittal view*